What
Safety
LEADERS
DO

The Insider's
Handbook for
**Safety
Leadership**
Tips, Tactics,
Secrets & Ideas

Matthew A. Forck

SAFE STRAT, LLC

ISBN Number 978-1475-13212-0

Printed in the United States of America

Published by

SafeStrat, LLC

Safety strategies . . . for LIFE!

Cover design, cover illustration, and book interior design by Susan Ferber, FerberDesign

This book is dedicated to all of those who work so hard to make a positive and powerful difference both in safety, and the world!

Contents

Introduction

There is probably no better path to safety success today than through strong, consistent, educated, and passionate leadership. At the same time, there is probably no other topic that is more misunderstood and complex than safety leadership.

The reason safety leadership is so difficult is at least two fold. First, to achieve long-term and sustainable safety success, an organization needs leadership at all levels. There is a need for strong leadership in the corner office, leadership from managers, leadership from supervisors and leadership from informal leaders—our workers who are turning the wrenches and getting the work done. Generally, we, as safety professionals, are trained for technical success. We spend hours in OSHA regulations, systems safety, fire systems, hazard analysis and the list goes on. Yet the key to long-term safety success, after our programs and procedures are technically solid and we have established a safety process, is our ability to get each of the above mentioned groups to lead. And not just lead, but to lead in an organized manner, in the same direction, and with the same goals and values . . . that's difficult!

The second difficult issue with leadership is there is no one path to leadership success—there is not a cookie-cutter template that we can implement that brings us to leadership effectiveness. Each organization must customize. This is different than many of our safety challenges. In most of our safety specific challenges, we can find one 'right' answer. We can find a reference book, an OSHA interpretation, a check list form or a subject matter expert to give us that one right response. But, since each organization has a little different culture, organizational structure, centralization versus decentralization philosophy, there is not one clear path forward for safety leadership—the path forward is cloudy, at best.

So, how can we put our organization in the best position to engage all leaders at all levels and build a custom leadership system for long-term sustainable results in the process? In short, capture leadership ideas, thoughts, concepts and insights and then apply the ones that fit in an organized and systematic fashion. That is exactly the opportunity this book provides!

Over the next few hundred pages find different concepts, ideas and mini templates on safety leadership. These diverse ideas are organized into different sections or concepts or mini-chapters, if you will. This book is organized so that any one chapter can 'stand on its own' for a specific concept. For example, The Safety Committee MAP (Monthly Action Plan): Our Course to Safety Committee Success is an independent chapter on how leaders can organize safety committees to be more effective. Or, Making Safety Top of Mind: Creating Awareness to Reduce Injuries and Near Miss Incidents, is a section on techniques safety leaders can use to increase safety awareness. Let these individual concepts support you in specific areas of need such as near miss reporting or supervisor effectives. Or read the book as a whole looking for trends and common themes—then harvest these common themes and apply them to your specific organization for best results.

Finally, you may get the best results from this book if you read and reread. Skip around and mark pages, then share them with your team and staff. This book is intended to be your leadership tool kit—tools that you can use to make your own customized leadership procedure . . . I wish you the best . . . now get leading!

Thanks!
—*Matt*

1 Managing the Desert Island

Ten Ideas When Safety Professionals Find Themselves All Alone

When it comes to safety in your organization, have you ever felt like it was you against the world? Recently, I received an email from a client. I could tell from the 'tone' of the email that she was in need of support. We arranged a conference call. It was on the call that she told me 'the story.' She was a safety professional for a medium sized manufacturer. Over the last few years, her attempt to move safety forward had been met on a number of levels with resistance. Her budget was thin. Support for ideas, products and programs was thinner. It was a 'check the box' culture, and no more. It was near the end of the conversation that I asked, "Do you feel like it's you against the world?" With a long sigh she quietly murmured, "Yes."

Unfortunately, she's not alone. Each day, safety professionals, safety supervisors, Human Resource staff with safety responsibilities and safety leaders feel like they are on a deserted island all alone. Even in organizations with supportive cultures, from time to time, one can feel this way. But, it's in times like these that safety professionals must re-double their efforts. Our environments are dangerous and mistakes can be life changing, not to mention the financial and legal exposures to our organizations. So instead of 'mailing it in' and waiting for a ship to come to your rescue, it is instead time to take bold action to advance safety and protect your people. Below are 10 ideas to help you take on the world when you feel all alone.

Create Buzz!—"Leaders are the stewards of organizational energy," writes Jim Loehr and Tony Schwartz in their groundbreaking book entitled, *The Power of Full Engagement*. Like it or not, you are a leader in your organization. Being a leader, you are accountable for your organization's energy. Chances are very good that if you are feeling alone the energy level within your organization is near rock bottom. At times like these, it's time to create buzz, with the purpose of pumping energy into your work group. There are a number of ways to do this from funny safety related YouTube videos to rubber chickens. I would suggest that you outline a formal strategy that includes a number of concepts such as raffles, give-a-ways and contests. Include every level, and make sure senior leaders are visible. This is not only good business; it's great safety and a lot of fun.

Play a Game—"When you can do the common things of life in an uncommon way," said George Washington Carver, "you will command the attention of the world." I know what you are thinking, there is no way that your people will 'play a game' or participate in an activity but let me tell you the secret, you don't tell them! There are tons of activities and games, that last anywhere from two minutes to more than a half hour, that your employees will do, not realizing it's a game or activity. I recommend purchasing a resource. I, of course, am partial to my book, *ISMA—Involved Safety Meeting Activities: 101 Ways to Get Your People Involved*. You can also checkout a helpful series called *Games Trainers Play*. Remember, engagement in an activity leads to involvement, involvement leads to ownership, and ownership leads to results . . . and it gets you off the island!

Give Something Away—Joe was a car salesman and he used some unorthodox techniques for results. For example, he deeply believed in the concept of handing an item to a prospective buyer. He explains in his book, *How to Sell Anything to Anybody,* that he had desk drawers full of items and that everyone that entered his office received a small token. He had a drawer of stuff for women. He had another desk drawer full of items for men. A third drawer was full of items for children. When people entered his office, the equation was unbalanced. In the customer's mind, Joe was going to be taking from them through a sale. In order to balance the equation, put the customer at ease, build trust and a relationship,

Joe would give each customer an item from his desk drawer, each customer, each time, no exceptions. We can use that same concept in safety. Give away a small token; in safety it is called a SAI (Safety Awareness Item). This can be a candy bar, stick of gum, duct tape or an apple. The item should be useful, practical and handed personally to the employee. Remember, this isn't about buying safety, it's about the relationship and putting the customer, I mean employee, at ease so you can have a good conversation and begin working your way off the island. Does this work? Well, if you don't believe me, believe Joe . . . he holds the world record for individual new car sales in one year, 1,425 and for a career, over 13,000 vehicles!

Block and Tackle—In sports, when things are not going well, the coach will always return to the fundamentals, or they get back to the 'blocking and tackling.' When we feel like it is us against the world we need to get back to safety's fundamentals; the safety process. A safety process is a formal and systematic set of procedures outlining safety in your organization. It includes management safety accountability, incident reporting, incident analysis, hazard recognition and inspections, and employee recognition, just to name a few. If you don't have a formal safety process, now would be the time to begin moving it forward. You can learn more from any number of reputable safety textbooks. If you have a process, take it out, dust it off and refresh your organization on blocking and tackling.

Share a Story—Joe is a safety professional and a close friend. For years we worked together in the safety trenches. It wasn't until I had known Joe for a number of years that he told me that a close friend of his had died right after college. His friend, who was more than seven foot tall, was just drafted into the NBA. Joe was supposed to meet this friend to go out and celebrate but his friend never showed. Joe learned later that night that his friend had taken a curve too fast and his car left the road. He wasn't wearing a seat belt and was thrown from the car and paralyzed. He died less than two years later from complications from the crash. Joe, for years, had been struggling with how to relate to our work groups when giving a safety meeting or training program. I looked him in the eye and said that you have to share that story, with everyone! Safety is personal and one of the best paths for organizational growth is through vulner-ability—the ability of people within the organization to share a story. To foster this, ask a number of people, especially senior leaders, to share a story. It doesn't matter what the story is, it matters that it is personal, heart-felt and honest. Keith Ferazzi, business leader and author, puts it like this, "Vulnerability is the courage to reveal your inner thoughts, warts and all, to another person." It's through this 'courage' that results are found.

Professional Development—There is an old saying that reads, "In order to help another, you must first help yourself." I was working as a line-manager for a mid-western utility when one day the phone rang. It was the safety manager. I thought he was calling about a near miss report, but to my surprise, he asked if I wanted to serve in a temporary roll on the company's safety staff. I accepted the opportunity and the position was made permanent a year later. I had earned my journey lineman card earlier in my career, so I knew the operational side of the business, but I didn't know safety. I began to work on my professional safety certification; I pursued the Certified Safety Professional (CSP) designa-tion. Nearly five years later, I passed my final exam! I tell this because through that process I learned safety. I found new ideas. I built great networks. I met study partners, who I still call on for support. If you find yourself feeling alone, don't forget to help yourself first. The new ideas and skills you find will help both you and your organization.

Find your Informal Safety Leaders—Leadership guru John Maxwell writes, "Leadership is influence over others." No matter how 'alone' you may feel, take a moment to look around your organization. What you will find is a handful of people who are leading. Each person will be leading in a unique and different way but each person will have influence over a number of fellow employees. Once you identify these leaders, bring them together and ask for their help. Engage them in activities. Use their leadership networks. Let this group talk to senior leaders, to share ideas and frustrations. Appreciate these leaders, for what they do. These are the people who can help you build a raft, and pull you from the desert island.

Bring Food—"The family that eats together stays together," reads that familiar saying, but why? Well, the family that eats together is not only sharing food but time and life. They are routinely together at the table talking about issues of the day, problems, and struggles. In sharing a meal, support is found. They are there, together, laughing and joking and telling stories. And, when problems arise they know, in part because of the time together at the table breaking bread, that someone is there to talk. It would be great to build that sort of atmosphere with a work group. Well, we can. Set up a structure to share a meal within your organization. For example, you can hold a monthly breakfast if the group makes it injury free or meets another safety related goal. You can hold quarterly luncheons. Set up a lunch series with workers and safety staff or better yet with workers and senior management. Over time you will find, the work group that eats together, is safe together!

Recognize the Power of Recognition—"Recognition is American's most underused motivational tool," said Richard M. Kovacevich CEO of Wells Fargo. The insightful book entitled, *The Invisible Employee* reported some remarkable statistics on the subject of recognition and appreciation. "According to a 2003 survey, 90% of workers say they want their leaders to notice their efforts and improve their recognition and rewards." In addition, "In an ongoing Gallup survey of more than 4 million employees worldwide, there is remarkable evidence of the business impact of recognition and praise. In a supporting analysis of 10,000 business units within 30 industries, Gallup found that employees who are recognized regularly increase their individual productivity, increase engagement among their colleagues, are more likely to stay with their organization, receive higher loyalty and satisfaction scores from customers, have better safety records and fewer accidents on the job." Safety is a tailor-made vehicle for appreciation and recognition. To work your way off the island, set up structured recognition including line and senior management. It is in the natural facilitation of safety process elements, such as job safety observations, safety team interaction, and safety goal achievements where consistent appreciation and recognition can grow roots and have a far-reaching positive affect on an organization.

Work One Day at a Time—Francis Petro, President and CEO of Hayes International Inc. said, "The fact is, the only day an employee can get injured is today. You can't get injured tomorrow until it gets here and you can't get injured yesterday because it is gone. So, we have to be very, very clearly focused on what is happening today and that becomes part of our makeup, that becomes part of our nature, and that becomes part of our culture." (McMillan 2007) Before 1997, Phillip Popovec, Site Director for International Specialty Products (ISP), said safety was, "terrible." But, the chemical manufacturer surmised, "We came to the conclusion that we don't have to worry about how many recordable injuries we get this year. We don't have to worry about how many recordable injuries we get this quarter. The only thing we have to worry about is not getting hurt today." (Smith 2005). Focusing on today allows everyone to clearly focus around today's hazard. It helps with job planning and job safety briefings. Focusing on today makes a clear theme that can bring energy back to your group . . . energy that can carry you gracefully off the island.

In the end, we all feel alone from time to time. It's not the fact that we feel that way, instead it is what we do about it that matters. Remember what Art Linkletter said, "Life turns out the best for those who make the best of the way life turns out." Keep fighting the good fight and working hard to be a desert island survivor.

2 Is Friendship Part of Leadership?

The Five Principles that No One Else is Talking About

In all of the leadership principles, strategies and studies published in the last decade, have we somehow lost one of the most important? Today, we have leadership lessons, guidance and training on all sorts of principles except one; friendship. Around 300 BC, Aristotle published a ten chapter book on leadership. Two of his ten chapters were dedicated to the leadership principle of friendship. Friendship was such an important part of leadership that it was the only subject that garnered two chapters. Today, however, as I stand in the library or my favorite bookstore, I can't find a leadership book focused on friendship; I can't find a leadership book with a chapter about friendship.

While the definition of friendship has changed with Facebook 'friends' and other social media interactions, the fact that it remains an important leadership tool has not changed. To that end, it is time once again to do what Aristotle did, bring friendship into the framework of leadership. Below are five leadership principles that no one is talking about that we should all be paying attention to.

Carry and Be Carried—Eric Grietens is a Rhodes Scholar, humanitarian and retired Navy SEAL. Upon returning from active duty in the Iraq and Afghanistan conflicts, Eric started "The Mission Continues," an organization that works with injured veterans, helping them get re-involved in the communities they left. Eric tells the story about a training exercise during the SEALs training. He says that at one point, they gathered with their 80 pound packs and were divided into groups of ten. The exercise was to jog ten miles in a specified amount of time with their heavy packs . . . and the kicker was that each SEAL had to be carried for one mile of the ten mile route. This training exercise, which teaches SEALs that they never leave one of their own behind, taught Eric that at some time we will have to carry someone on our team, and at other points we will have to be carried.

Leadership is understanding this equation . . . that in our personal lives and our work worlds we must be willing to carry others and at times be carried ourselves. Today's society and culture is generally very intolerant of those who can't 'carry' their own weight. Leadership is knowing when to carry and when to be carried . . . doing it well is also called friendship.

Be Vulnerable—After college, I accepted a job that the college guidance counselor never talked about, meter reader at the local utility! A short time later I was promoted to apprentice line worker. Electric line work is a very rewarding job. You work with your hands and ultimately supply one of the most important needs of today's society, electricity. Shortly after I was fully qualified as a 'journey line worker' a big October wind storm hit. Due to massive power outages, I was called out to help restore power. My first assignment that night was to go to a substation. I needed to switch the electricity so crews could repair a line. In the wind and driving rain, I made a switching error, and in so doing, ignited a large electrical fire in the substation. After watching it burn for what seemed like years, but was probably only a few seconds, back up sensors put the fire out.

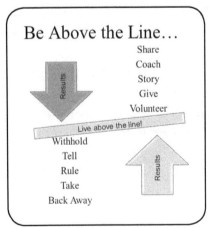

"Vulnerability," Keith Ferazzi, business leader and author says, "Is the courage to reveal your inner thoughts, warts and all, to another person." I think the interesting thing about this story, in addition to the fact that I did not get fired, is that I never shared it. A couple of years after it happened I was promoted to a line-manager. And, a few years after that I moved into the utility's safety staff. In this story there are some great lessons for others to learn from, but I didn't share because I was afraid to be vulnerable—I was afraid to live above the line.

The chart to the right is a guide to vulnerability. In any given day or on any given circumstance, are you giving, sharing and coaching, or, telling, taking and withholding? Vulnerability is the portal to engagement and ownership, which equals results! Vulnerability is also the first step in friendship.

Foster Friendship in the Workplace for a Competitive Advantage—It's true that when we think of a friend, we generally think of someone outside the workplace. Yet, we spend more time with our co-workers than our friends . . . we spend more time with our co-workers than anyone else! That being said, why not make a friend or two at work?

In fact, the *Gallop Management Journal,* the publishing wing of the Gallop Polling Group, said this, "When it comes to innovation, business leaders aren't necessarily looking to traditional sources, like research and development departments, to contribute big new ideas. Rather, they're counting on ideas from their employees, customers, and partners to help drive the organization forward. And engaged employees are most likely to contribute those innovations." They went on to write, "Researchers also explored the role that workplace friendships play in promoting innovation. About three-fourths of engaged employees (76%) strongly agreed with the statement "I have a friend at work who I share new ideas with." On the other hand, only 2 in 10 actively disengaged employees (21%) strongly agreed that they have a friend at work with whom they share new ideas. Clearly, friendships do play a significant role among engaged employees when it comes to setting the stage for idea creation and refinement."

What friendships do we have at work? How can we foster those relationships? And, as managers and supervisors, how can we foster friendships among our staff and teams? Doing this can give you a friend . . . and a competitive advantage.

Everything Moves Through Networks—One of the most horrid human tragedies of the last century happened in Rwanda. It started on April 6, 1994 when the Hutu led Interahamwe began to slaughter the Tutsi population. Within every story of human loss, there is a story of human hope, and this one is no different. As the Hutu led ethnic cleansing began to unfold Paul Rusesabagina began to use his networks to take action. Rusesabagina was an assistant hotel manager for the Mille Collines hotel. After the conflict began, the manager and most of the hotel staff fled the country, but not Rusesabagina. He used his connections within the hotel front office to be named manager. That gave him the opportunity to begin bringing in refugees, to hide and house in the hotel, until the conflict was over. Rusesabagina and those he was trying to save would not have made it without Rusesabagina's network. He knew everyone and many were willing to help, from food to other supplies.

Today, everyone knows the importance of a strong network but few of us act like we know. Instead, many of us take our networks for granted, until we need them. To strengthen your network, spend five minutes everyday sending two or three emails to someone in your network. Send them an interesting article or simply touch base. Or, instead of scarfing down your lunch at your desk, set a goal to eat lunch at least twice a week with people from your network. Fostering these resources is one of the best leadership investments you can make. If you don't believe me just ask Paul Rusesabagina. He, through his networks, was able to keep Hotel Rwanda open and save 1,268 people in the process.

Energy—There is an old story that goes like this, "There was a man walking through town and he saw a buddy over on the bridge getting ready to jump. The man ran over and encouraged his friend to come down and talk. He did come down off the bridge and they did talk . . . and two hours later they both jumped!"

The point is simple, no one likes to be around an energy drain. As a matter of fact, it's just the opposite; we need to bring energy into each day, each situation, each interaction. In so doing, we are fostering that relationship, that friendship. Jim Loehr and Tony Schwartz in their book titled, *The Power of Full Engagement; Managing Energy, Not Time, is the key to High Performance and Person Renewal* shared the following three concepts in their book. First, "Energy, not time, is the fundamental currency of high performance. Performance, health and happiness are grounded in the skillful management of energy." And finally, "Leaders are the stewards of organizational energy!" Do we save our energy for after hours, or do we bring it everywhere we go, including work? Energy helps make and foster friendships!

"Friendship," Aristotle said, "is a single soul dwelling in two bodies." Friendship is a powerful leadership tool. A tool that no one is talking about, except leaders.

3 Safety Isn't Number One!

How Safety Leaders Work with the Corner Office

Take the following quiz about your person in the corner office. In the last ten presentations given by your CEO, what was the main topic? Next, if you looked at the agenda presented at your company's most recent board of director meeting, what would be the primary focus? Finally, if you looked at your CEO's calendar this week, what is the purpose or topic in the majority of his/her meetings?

I have a good friend who teaches for a MBA program. To begin the session he will give his students a 10 page business scenario. Along with a description of the business, the packet includes balance sheets, bank statements, revenue flows and production costs. The students have a few days to read the material and come to class ready to discuss.

My friend tells me that each semester the discussion is very similar. When he asks the class what the major problem is with the business, the responses are varied. Someone on the back row will immediately speak up and suggest that the company has a problem because of a very weak social media presence. Another will point to a slow production process compared to their competition. One will propose that the company has branding and customer identity issues. Yet another will submit that the company is struggling because it has not tapped overseas markets. Finally, after nearly two hours of debate a quiet person on the front row will timidly raise her hand and say, "I'm not sure it's what you are looking for, but the way I read the cash flows, this company isn't going to make payroll next week."

My friend will then hurry to the board and write in enormous letters, C-A-S-H. "Cash, my students, is the most important thing in business . . . never forget that."

If you have taken the quiz above you have discovered what your CEO and senior leaders already know; cash is the most important thing in business. Cash sometimes goes by different names in different businesses like dividends, earnings per share, stockholder value, budget compliance, profit margin or simply making payroll. My guess is that your CEO and senior leaders manage cash very well. Because they are so good at it, they were promoted—that is the reason they are now leading your company! Senior leaders and CEOs are laser focused on cash, no matter how they say it. And we should want it that way, right?

Unclear Expectations by Safety Professionals—So, what's the problem? In the last decade safety professionals have consulted, written, urged and established that an organization can only be effective if these same cash driven senior leaders make safety a value. Not only must they make it a value, they must get out there and support it! So, our CEO will inadvertently end up at an employee meeting where safety comes up. He will then articulate that it's the most important thing. Thirty minutes later he will be back in his office making a cash related decision that in the eyes of the employees appears to contradict his statements on safety.

In truth, we more than likely, have set him and our organizations up to fail. Failure comes in a number of ways. Failure means the employees believe there is a 'say-do' gap on the part of the CEO and senior management, because they say one thing about safety and then do another. Failure is the appearance that senior management team is disconnected. Or, that it is 'wrongly connected;' meaning they want to help and be supportive but have not been given clear direction. In short, we do need our CEO and senior management team's support in order to run an effective safety program. And, in order to get their support, we as safety professionals need to be very clear about what 'support' looks like. Here are five things that safety professionals should do with their CEO and senior management.

Give Your CEO Three Things To Do—"One of safety's greatest failings has been its total inability to involve top management in safety," Dan Peterson writes in Safety By Objectives. "Probably all corporate presidents in this country want safety for their people and yet they usually fail to do anything to achieve it. This is more often than not the fault of the safety managers. We simply have not told these presidents what it is that they must do to achieve safety. Safety managers have traditionally bemoaned the fact that they need management support and cannot get it."

In working with our CEO's, we must give them specific things to do. I suggest three items each month. I like one written, like a note to a supervisor or field worker related to safety, for example. Think about the ripple effect this has on your organization. Scribbling a note takes no more than 15 minutes. Consider one spoken, like discussing safety with a small group of workers. In this setting, have your CEO share a personal story; it will make him or her seem more human and approachable. This takes about 30 minutes. Finally, schedule a field related visit. Have a supervisor or safety professional accompany your CEO to the field or through the shop floor. Stop and talk safety with as many workers as you can. This activity, depending on your industry can take anywhere from an hour to four hours. In total, you are asking about five hours or less of your CEO's time per month . . . and your expectations are clear.

Using the Calendar to Your Advantage—While the goal of your CEO is cash, he or she is driven by their calendar. As you know, CEOs are incredibly busy and each day to them presents ten to twelve hours of appointments and meetings. But, if we understand that CEOs are driven by their calendars, we can plug into those calendars to help us with safety. In addition to giving your CEO three specific things to do, schedule a monthly or bi-monthly meeting with him/her. This doesn't have to be a long meeting, thirty minutes to an hour will be sufficient. This meeting has many purposes. First, it gives you a chance to better know and understand the CEO and gives the CEO a better chance to know you, and safety. You can share status updates on safety goals and discuss any recent incidents. In short, just as the CEO meets with other critical department heads, you need to be on that rotation, educating and informing the person in the corner office.

What is the Score?—"You play different," the old saying reads, "when you know the score." One of the primary responsibilities is to make sure your CEIO and senior team know the 'safety score.' You can bet that your CEO gets daily or weekly 'scores,' sometimes called 'dashboard' reports, related to cash. These are typically one page, real time reports that offer a quick snap shot on the exact financial health of the company. What kind of scores do we offer? Traditionally, we offer a once a month injury report. And, this report generally comes out about the middle of the following month, so an injury could be more than 45 days past before making it to the CEO's desk. What if we initiated a safety dashboard? Information could include injuries, the cost of injuries, near miss reports, job observations, safety audit scores, and more. This real time report is typically much different than the reports that are currently provided. But, setting up a process to give our CEO a real time score (safety dashboard) will not only give you and your safety team real time data to identify trends, but it will leave a positive impression on your CEO as well.

Appoint a Vicar—Now that you are on your CEO's calendar on a regular bases, you have worked with your CEO on three, or so, specific activities and you are giving him/her a weekly dashboards, it's time to take the next step. Just because your CEO can't be out in front of work groups talking safety day in and day out doesn't mean that it's not a very important role for a member of your senior management team. So, who is your safety vicar? If you have one great, if not have your CEO appoint one. A vicar is a representative who is entrusted to act "in the person of" or agent of the CEO. In this case, it is to carry out the CEO's message about safety. It is best if this person is an operations manager or VP and has 'the ear' of the CEO on a regular basis. There needs to be close communication and goal alignment between the safety staff, the vicar and the CEO. This is a very good way to have a safety champion out with your workers day in and day out.

Strive for Operational Excellence—Finally, make safety part of operational excellence instead of making operational excellence about safety. Many organizations fall into the trap of making safety about 'the right thing to do' or a moral issue. I agree that it is both, but if we want to be more effect we need to be able to build a solid business case around safety; a business case that has a strong and sound financial foundation. To make this case, we should strive for operational excellence. Operational excellence is striving to perform in all aspects of your business, from product quality to human performance, to safety performance—they are all tied together. The more we can talk about operational excellence and convince our CEO that operational excellence is the threshold to exceptional safety performance, the more successful we and our organizations will be.

In the end, we can't be successful without the person in the corner office. Just because safety isn't the most important thing on your CEO's mind doesn't mean that you and he/she can't forge a strong bond. It is a relationship that can move your organization forward toward operational excellence—and safety!

4 Not Another Safety Meeting!

Making Sense and Getting Results Out of Safety Meetings

"I have heard many sermons," a man once said about attending church, "and I have even listened to a few!" While this may or may not hold true for church, it certainly holds true for safety meetings. I have witnessed it time after time; our workers drag themselves into the weekly safety meeting, sit in the same place, hear the same person talk on a topic, then go to work. All the while, the Manager is looking on wondering what return she is getting for the investment. Our people may be listening, but more times than not, they do not hear!

'Safety Meetings' are as traditional as incident reviews, OSHA, Safety Committees, Safety rule books and posters! Safety meetings are one of the foundational elements to a solid safety program. Weekly safety meetings have many purposes, all rolled into that 30-minute agenda. First, they are a communication tool, to discuss any changes or hazards needing 'special attention.' The time can be used for training, motivation, change management, fun and community. That's what they were supposed to be, anyway. The problem is that over the last few decades, safety meetings have lost their edge. Meeting leaders are out of ideas. Meetings lack preparation, planning, and a clear agenda. Participants have disengaged. Meeting results have suffered—until now.

The following is a systematic approach to the weekly safety meeting. It's is an outline for 52 safety meetings. This outline is designed to be a mix of solid safety rules and procedure, interaction, energy and upper management support and communication. Let's get started! In the course of one year, 52 safety meetings, I would recommend:

16 Rules/Procedure-Driven Training Sessions—"The training regime, however, were exhausting on every level of human effort, for both astronauts and their ground controllers," writes Craig Nelson in his epic book called *Rocket Men,* the story of Apollo 11. "After learning the basics on a sim-flight that went smoothly, both sides were subsequently put through their paces with a series of problems like mechanical break down, conflicting data streams and all out system failures." Nelson writes later in the book about simulation testing, "This process worked so well, in fact, that in time many astronauts would calm themselves in real work crisis by thinking, *this is just like simulation.*" Because of their training, they had great confidence!

If seems only 'right' that the majority, 16, of the weekly safety meetings would be spent on mitigating hazards through training procedures—the things that matter most. Holding training or safety rule reviews in weekly meetings is not intended to replace current training schedules or modules. Instead, holding refresher training during these weekly meetings will raise awareness on key hazards moments before your people begin their work for the day. The real question is, are we training to 'check the box' and comply with some annual federal requirement or do we train like the astronauts of Apollo 11, as if our lives depend on it! Here are some ideas to raise the urgency and engagement around training. Have an employee share a personal story of an injury or near miss as an introduction to the training. Simulate situations, just like the astronaut's by shifting training from sit and listen to get up and do. Have your employees add 'system failures' to training and discuss how they would work around them safely. Test for comprehension, holding employees accountable for test scores. Be prepared and be original.

12 Heart to Head Involved Safety Meeting Activities—Over the last two decades, Duke University Men's Basketball team has had as much success, if not more, than any other team in all of sports. To find the secrets to their success, one has to look no further than their Coach Mike Krzyzewski. Given the fact Coach Krzyzewski mentors and coaches 18 to 22-year-old men, how many team rules do you think he has? Probably at least 25? Maybe more than 100? No, Coach Krzyzewski has **one** team rule, "Don't do something detrimental to yourself." In hurting yourself you are hurting the team and the school too.

We traditionally solve organizational problems with a new rule. In fact, the commonly used tools for our employees are rules, strategies, policies, structures, procedures, monitory awards and discipline. These techniques are the outside-in approach and they yield very limited results. If lasting success and sustainable results are the goal, organizations must first go to the heart before engaging the 'head'. The heart deals with beliefs, habits, energy, passion, personal commitment and personal goals. When the heart is engaged, results will follow! Or as Coach Krzyzewski says, "Don't do something detrimental to yourself!"

To be successful with 'Heart to Head' interaction, invest in some resources such as *ISMA—Involved Safety Meeting Activities: 101 Ways to Get Your People Involved*. Or search the internet with key words, 'games trainers play.' Once you have ideas, be prepared. Match meeting themes to seasons. Remember, the key is to get people thinking about what matters most, and then get them involved, have fun and allow them to share.

12 Outside Speakers—Several years ago when I was working as a safety professional for an utility company, I served most of out-state Missouri. Within my territory was one work location in a town with the population of just over 60,000. The supervisor, who organized the safety meetings, had a goal to bring in one outside speaker per month. He defined an outside speaker as someone outside of the immediate work group. The speaker could be from another part of the company or from outside the company altogether. I first thought this was a lofty goal, one that couldn't be met. Yet, year after year, I saw outside speakers come to safety meetings, one per month. I saw conservation agents, doctors, police chiefs, local coaches, eagle scouts, OSHA reps., managers from other plants, safety professionals from other companies, industry experts, etc. What I really saw were results!

The bottom line is his people listened to these outside speakers. They were new, different, fresh, entertaining and had a story to tell that no one had heard. They key is to organize a small sub-committee for outside speakers. Brainstorm a list of potential candidates. Guide and coach the speaker about what you do and how he/she can help with your overall safety efforts. Tie the outside speaker's message back to your hazards and your safety. And . . . get results!

Two Professional Speakers—Why wouldn't we, twice a year, bring in a professional speaker to make our people laugh, cry, interact and learn more about critical concepts such as team building, leadership, accountability, responsibility? One word-cost! But let's really look at cost.

Let's say for example, that your employees make $20 per hour (many trades and skilled craft are twice this amount) and the employees benefit package is another $20 per hour. Then it costs you $20 per employee for a half hour safety meeting. But wait, production, tools, materials and overhead are often at least a 5:1 ratio of salary. So, when adding that, it costs you $100 per employee for every half hour safety meeting! But wait one more minute, Louis J. DiBerardinis in his book, *Handbook of Occupational Safety and Health* writes, "The average direct and indirect cost of an employee injury accident is $20,000. This translates into the company having to generate $200,000 in sales/services at a 10% profit margin, or $400,000 at a 5% profit margin, to pay for a $20,000 accident."

In short, the cost of a professional speaker is pennies compared to the cost of a weekly safety meeting and/or injury—the real question is, can you afford not to? When looking at professional speakers get a short list of potential speakers then request a presentation package from each. Check references. Expect a presentation outline from the speaker so that the speakers message is consistent with your objectives. Finally, work with the speaker to identify key themes, tools and catch phrases so that the momentum from the presentation will continue for weeks and months.

Two Appreciation Days—I have never received too many phone calls from my boss telling me that I am doing a great job! I am yet to talk to someone who tells me they are over appreciated at work or at home. Carve out two safety meetings a year to appreciate your work group. This can take many forms and you may want to consider a sub-committee who takes the lead to plan these meetings. Whether it is an end-of-the-day meeting with ice cream, a fun awards luncheon or pies in the manager's face, appreciate your people!

Four Safety Updates and/or Incident Reviews—What is the most common injury on our property? What day of the week are we most likely to get hurt? Are more people injured before or after vacation? What were the key recommendations from the last near miss or incident summary? These are facts, figures and data on the desk of safety staff members and managers but it is all information that needs to be on the minds of our workers. Knowledge coupled with awareness is so very important and can lead to injury reduction. These safety meetings, once per quarter, are must-dos.

Four Upper Manager Reviews—It doesn't matter where I am speaking, or to what industry, if I ask what is the biggest 'knock' against our upper management in terms of safety, the response is the same; lack of feedback and face time! To that end, plug your manager, vice president, or other senior leaders, into your safety meeting schedule on a quarterly basis. When these leaders are in front of your group, coach them to be vulnerable. Keith Ferrazzi, business leader and author, defines vulnerability as "The courage to reveal your inner thoughts, warts and all, to another person." If your senior leaders can take that risk, around an incident in their garage, an experience with one of their children or an injury when they were younger, that will go a long way in creating the trust and credibility needed for long-term success.

Well, now you have it, 52 safety meetings—an outline for an entire year. Yes, it's a lot to think about, a year of safety meetings, but remember you are not alone. In the coming weeks put together a sub-committees, as needed, to support this effort. Meet with your safety and training staff to discuss. Talk to your senior leadership about their role. Safety meetings are a foundational part of safety, but they don't have to be 'traditional.' Remember, what you invest is what you will get back!

5 Safety's W.I.L.D. Thing

What It Takes to be a Successful Safety Team

Lou Brock's career stats are off the charts. In his 18 year major league career, Lou batted nearly .300, had 900 runs batted in and stole 938 bases. But in terms of baseball stats, Lou is most proud of what he wears on his fingers. There are only five players in Major League baseball history that have five or more World Series rings . . . Lou Brock is one of those players! The point is that Lou knows how to win. In a recent interview, Lou said that each championship team must have a wild card; a player who possesses energy, who pushes positive limits, and talks to everyone on the roster. This player may not be the most talented in the field but he/she brings a spirit and enthusiasm that makes the whole team better.

I thought of what Lou said and realized that this 'wild card' must be present on our safety teams too. The only difference, I think, is that in safety each and every person must be W.I.L.D. What I mean is that in safety sensitive work members of the team need to be W- willing to talk. I- identify the plan, L- possesses lively energy and D- determined to make a difference. Let me explain.

W: *Willing to Talk*—Mary Kramer and her family had a new neighbor, Bella. Bella wouldn't have been their pick of neighbor's but early on they decided to tolerate the pit bull puppy. As the dog grew, it never really caused trouble. Sure it would eat her dog's food and chase her horses, but what could she do. In Boone County, Missouri there wasn't a leash law for dogs as long as one was out of 'city' limits. Mary thought about talking to the owners but stopped short saying, "You never know how somebody's going to take it. I don't want to be confrontational."

Last week Mary's ten-year-old son Tyler was walking to the bus stop. From behind he heard the sound of 'steps on gravel'. He turned to see Bella in mid-air lunging toward him. Tyler knew that if attacked by a dog or other large animal that one should curl up in the fetal position guarding the head and neck. That's what Tyler did, he forgot however about the last half of that instruction, to remain quiet, if possible.

Curled up, screaming, Tyler laid in the grass of the home next door as Bella ripped his backpack and jeans to shreds. Mary heard her son's screams and immediately came running outside. Seeing her son being attacked was all that she could endure. A child being attacked by a pit bull is life flashing in front of one's eyes, a brick firmly planted in the center of the gut and adrenaline, a lot of adrenaline. Mary charged Bella as he chewed her son's backpack. Luckily the dog ran off. Mary picked up Tyler returning to the safety of their home to call authorities.

The old wise quote from an anonymous author says, "An excuse is just a reason packaged with a lie." And that is exactly what we tell ourselves when we don't speak out or talk about safety. If we are going to be WILD about safety, the first step is for us to be willing to talk . . . not worrying how someone is going to 'take it.'

I: *Identify the Plan*—What images come to mind when the words 'family vacation' are uttered? How about the mention of an 'European family vacation?' On Monday, August 4, 2008, that is exactly where an Israeli couple was taking their family . . . all five children! Since it was Monday, nothing was going right. They had trouble getting their children and 18 suitcases into the vehicle so grandma could shuttle them to Ben Gurion airport. After all, they had a deal with the airline company; if they weren't on-board at flight time, the plane would leave them behind!

As they arrived at Ben Gurion airport, they quickly realized why it is the busiest airport in Israel. Being late, they didn't have time to think about the crowds of people and all of the confusion these crowds cause; instead they had to get their bags and children to the terminal, fast. The plane to Paris would leave if they weren't on it, so they ran fast, as fast as one can in a crowded international airport with 18 pieces of luggage, five children and thousands of other travelers doing the same in all different directions. As they relaxed in their seats and the plane taxied down the runway, they were able to take a deep breath, they had made it, yet something seemed to be missing.

Back at the airport, a very cute little girl pulled on the pant leg of a police officer. The officer squatted in order to look the toddler in the eye. The girl said, "Where are my parents, where are my brothers?' The officer quickly learned that the family had just boarded a flight to Paris, and somehow left their child behind. The officer had about 120-seconds to get the girl to the terminal before the plane would be gone. She grabbed the girl and ran, but it was too late. They both watched the jet fly into the clouds. The parents, after being in the air for 40-minutes were finally notified that their daughter was safe with authorities at the airport. The parents had not yet realized they were missing a child!

"Men in the game are blind to what men looking on see clearly," a wise Chinese proverb reads. And, if we jump into the game, (our work,) without making a plan, we are blind to hazards that can get us hurt. Being WILD about safety means that we plan first, then work the plan. If you don't believe me, listen to this little girl's grandmother. "We're in shock. They're very responsible and organized, top-notch people." Top notched people in a hurry without a plan.

L: *Lively Energy*—Most scientists have dreams of that 'big break through,' and a set of goals to make that dream a reality. For David Atkinson, it was no different. David, a scientist and astronomer working at the University of Idaho, had a goal to study the wind patterns on the moon. No, not the earth's moon; his goal was bigger and better. He wanted to study Saturn's largest moon . . . Titan.

After setting the goal, David and his research team went to work. They planned, diagramed and plotted. They studied, mapped and engineered. The only problem was that NASA didn't have any plans to launch on the ringed planet. David was patient. Time went by . . . actually a decade went by. David and his team continued to refine their plan, submit proposals to NASA and waited. After eighteen long years in a quest to discover the secrets of wind activity on Titan, David received word that a European Space Agency was planning a mission to Saturn. He immediately contacted them and submitted his work. His dream would come true; they agreed to allow a probe to land on Titan.

The probe would land through David's hard work and through a joint venture between the Italian Space agency and NASA. Once the probe landed, David and his team would study the wind movement through a radio wave from the probe to their lab back on earth. The probe was equipped with two channels. Channel-A was the primary channel, powerful enough to send the signal clearly back to earth. Channel-B was a much weaker back up channel, for emergencies only, and transmissions through this channel would take years to decipher.

The probe was launched and nearly two years later it landed. David and his team huddled in the lab to receive the radio-wave signals and begin study . . . nothing. They waited some more . . . nothing. They ran tests to try to figure out why channel A was not sending a signal; nothing. David never received a signal, as he learned that someone forgot to turn on channel-A!

"In order to be burned out, you have to first be on fire," an old saying reads. Are we 'on fire' about safety. Do we exert the energy that our safety deserves when it comes to job planning, safety inspections, safety meetings, safety committees, or do we forget to turn on our energy? "I and the rest of my team waited and waited and waited," Atkinson later wrote. "We watched the probe enter and start transmitting data but our instruments never turned on. We do have channel B data and although driven by a very poor and unstable oscillator, we may be able to get a little bit of data. I think the key lesson is this; if you are looking for a job with instant and guaranteed success, this isn't it." Turn your energy on today . . . before it's too late.

D: *Determined to Make a Difference . . . In Our Own Way!*—Cliff Young strolled to the registration table and asked if he could participate in the race scheduled to begin early the next morning. "You'd have to wear running clothes and jogging shoes," the register suggested. After Cliff explained that he preferred his coveralls and gum boots, he paid his entry fee and was officially registered for the 1983 Sydney to Melbourne Ultra Marathon.

By definition, an ultra marathon is anything longer than the traditional twenty-six plus mile marathon. I don't know about you; but when I'm at the gym, my favorite machine is the vending machine. A twenty-six mile marathon would wear me out if I drove it, let alone ran; I can't imagine one that would go farther. And farther the Sydney–Melbourne race went, depending on year and exact course it was between 810 and 1100 kilometers.

The next morning Cliff was ready to race, sporting what he was comfortable in, coveralls and gumboots. Since he didn't have any qualifying times from other races to present the registrar, he had to begin at the end of the starting line. The world-class ultra marathoners were at the front of the starting line, the middle-of-the-roaders next, followed by those unranked competitors . . . Cliff among them. He figured that over a several day race starting at the end of the starting line would not matter. The starting gun sounded. The crowd began to move, and in mere minutes Cliff lost sight of the world-class runners . . . they were too far ahead.

This race, run over the course of several days, is generally run in packs. The world-class athletes will stay together and run for about eighteen hours straight. Then stop, rest and sleep for about six hours, doing it all over the next day. Once they are about three to five miles from the finish line, the pace picks up; eventually one runner will out sprint the others to take the ribbon.

That is what happened in 1983 as well. After several days of running for a dozen and a half hours and sleeping for six, the pack of world-class runners broke out into a sprint. Finally one pulled ahead and crossed the finish line to the cheer of the crowd. He fell to his knees exhausted from the grueling event. As he was greeted by a race official the runner looked up and mustered a few words.

"I thought you guys would have put a ribbon across the finish line," he said as he gasped for breath. "We did," promptly responded the race official. "I know I have been running for days, but I still know a ribbon when I see one, and there was no ribbon."

"You didn't finish first."

"That can't be . . . I was in front of all of the world class group; who beat me?"

"I can't remember his name," said the race official, "some guy with coveralls and gumboots."

And so it was that Cliff Young not only won the 1983 Sydney to Melbourne Ultra Marathon, Cliff Young set a new race record, beating the old mark by nearly two days. You see, Cliff had never participated in a race like this. He didn't know about the 'run 18, rest 6 rule.' Instead he thought it best if he ran 22 hours and rested 2. So, sometime in the dark of the third night, Cliff Young coolly ran past the world-class runners as they slept.

Only you can do the work that you have been sent here to do. The key is that we must do it our way, just as Cliff did. By the way, how old do you think Cliff Young was in 1983? Cliff Young set this record at the age of 63. It's never too late, lead your way!

In the end, every team needs a WILD thing, and safety is no different. The only difference is that on our safety teams, we all must be a little WILD, if we are going to win, finish safe, each and every day!

6 What is Your Safety Vision?

Some Thoughts About Where Safe Needs to Go in the New Year and Beyond!

A good friend called a couple of weeks ago. He was both excited and nervous. His excitement was due to the fact that he was interviewing for a safety manager position on a corporate safety team. If successful, he would be leading a team with over a half dozen direct reports and influence throughout a nearly 10,000 person organization. He was nervous, not because of the interview, but because he wanted to write a comprehensive safety vision to present in the interview; and he had only a couple of days to pull it together. He called me to pick my brain and pull together some ideas about where safety should go in the next year and beyond. This is the safety vision we put together.

Leadership is No Longer About Senior Management. Over the last five years one couldn't throw a dart at a safety magazine without hitting an article on leadership. And, probably 99 out of 100 of those articles were about the vital role of CEOs and senior leaders in the safety process. During this time, an equally important group of safety leaders has been working 'behind the curtains' to affect positive change. This was proven in a recent survey, *"A recent survey of a company with 2,500 employees asked the employees to identify leaders within their organization. Only 15% of the identified leaders had a title of manager."*

Each organization has a group of leaders, called, informal safety leaders. They have no official title or rank but possess great influence in our organizations. The key to safety success in the next year and the next decade will be the ability to identify these informal safety leaders, align them around the organizations safety goals and beliefs, then engage them in a meaningful fashion. Senior leaders are still important, but true success will come through the relationship with the informal safety leaders, the ones with true influence over the everyday choices and actions of our people.

Safety—the New Measure of Performance. About two years ago, a safety professional leading a major utility company gave a presentation. I apologize that I do not remember her name, but I did write down what I believe is a key statement that she made during the presentation. "Some investors," she began, "simply look at a company's safety performance numbers, because if a company can manage safety, they can manage their business and will be a solid investment." Safety is a unique mix of managing people, attitudes, training, behavior, tools, systems, policies and procedures. It is not a performance measure in the next year, but it is 'the' performance measure in the next decade! Companies that can manage safety will not only be ahead of their peers in an OSHA reportable data but will experience a competitive advantage in a number of areas such as human performance, workers compensation and medical costs, employee engagement and worker turnover. Safety is not only the right thing to do, but also it's the right business thing to do, and moving forward, it is not just safety professionals that will be noticing.

A Critical Shift from Head to Heart. Duke University Men's Basketball team has had as much, if not more, success over the last two decades than any other team in all of sports. To find the secrets to their success, one has to look no further than their Coach Mike Krzyzewski. Given the fact that Coach Krzyzewski mentors and coaches 18 to 22-year-old student athletes, he probably has to have an entire rule book to keep them in line and produce results. To that end, how many team rules do you think he has? At least 25? Maybe more than 100? No, Coach Krzyzewski has one team rule and that team rule is simply, "Don't do something detrimental to yourself." In hurting yourself you are hurting the team, and the school too.

It's a trap that every organization falls into. A deficiency is found so a new rule is established. Along with that new rule is a training session, a safety meeting, a set of posters and a lot of 'buzz.' After six months; however, the results are the same. Because there was no improvement in results, a more stringent rule is established and the circle completes again, and again and again. The problem is that rules, strategies, policies, structures, procedures, monetary awards and discipline all deal in the head. It's the outside-in approach to behavior and it yields very limited results. Sure policies, rules, discipline and accountability must be in place, but that isn't a ticket to results, it is just the foundation. Moving forward, lasting success and sustainable results will be found when organizations go to the heart. Going to the heart means that we are dealing with beliefs, habits, energy, passion, personal commitment and personal goals. It means involved safety meetings, trusted coaching and feedback, and interactive and real time near miss programs. When the heart is engaged, results will follow! Or as Coach Krzyzewski says, "don't do something detrimental to yourself!"

Finally, there is a new definition of Win-win. In the past, it was 'us' against 'them.' It didn't matter if it was 'safety' versus 'operations' or 'management' versus 'labor' or 'senior leaders' against 'middle management,' there was always tension. Moving forward, there is no scenario where management wins and the labor/unions lose. There isn't a model where safety staff wins and line management loses—we are all in this together. If you don't believe me, ask a few of our automakers! Organizations who understand, from top to bottom, that everyone is on the same ship and if the ship sinks we all sink, will have a competitive advantage, one that is tailor made for safety results. In this true win-win environment, teams will take on new meaning and new energy. In a true win-win environment, we assume innocence and trust. Organizations who understand this relationship and promote this type of environment, will rewrite the definition of 'team.'

In the end, my friend earned the job. He assumed his new role on January 1 of this year and is now working to bring his safety vision to life . . . what is your safety vision, and what are you doing to make it a reality? Happy New Year!

7 Lessons from a Legend
What John Wooden
can Teach About Safety . . .

On June 4th, the world lost a legend in basketball as Hall of Fame player and coach John Wooden died at the age of 99. Wooden, born on October 14, 1910 grew up in rural Indiana. His love for basketball grew with him, and he led his high school team to the state final three years in a row, winning the championship in 1927. He attended college at Purdue, helping his university win the 1932 National title. While at Purdue, he was the first player ever to earn three consecutive All American honors. He met Nellie (Nell) Riley in 1926 and they married in 1932. He earned a degree in English and worked as a teacher immediately after college. He also played professional basketball for the Indianapolis Kautskys. In 1942 he joined the Navy to support his country and fight in World War II. There he served for nearly three years. After his military service, Wooden taught and coached at Indiana State University for two years, before accepting a job with a struggling basketball program in California, UCLA. There, he and his teams made history. From 1948 to 1974, Wooden's UCLA Bruins won 620 games and 10 national titles. His teams had four perfect 30–0 seasons, won 38 straight NCAA tournament games, held an 88 game winning streak and won 98 straight home games. With this much success, both as a person and a coach, there is much that safety professionals, managers and leaders can learn from John Wooden.

"Sports do not build character. They reveal it," John Wooden. Coach Wooden required players to keep their hair neatly cut and he did not allow any facial hair. Legend has it that UCLA's All-American Center Bill Walton reported to the first practice his senior year with a beard. When Coach Wooden saw him, he asked what that was on his face. Walton, with all the courage he could muster mumbled, "Coach, it's a beard and I believe that we should have the right to wear a beard if we want." Loosing an All-American center like Walton would have probably ended the streak of national title runs for Wooden and UCLA but coach Wooden never missed a beat. "Bill," Wooden quickly replied, "I agree, you should be able to wear a beard if you want . . . and we will miss you on this team." Walton returned to practice five minutes later, clean-shaven! Safety, like sports, reveals character. It doesn't matter if the production line is down, the power is off, or orders are going through the roof, there is always the opportunity it cut a corner and there is always the opportunity to do it right . . . what do you chose? What is your character?

"Be more concerned with your character than your reputation.
Your character is what you really are, while your reputation is merely what
others think you are."

—John Wooden

Sweat the Small Stuff—"You know, basketball is a game that's played on a hardwood floor," Wooden said. "And to be good, you have to . . . change your direction, change your pace. That's hard on your feet. Your feet are very important. And if you don't have every wrinkle out of your sock . . . "Wooden continued, "Now pull it up in the back, pull it up real good, real strong. Now run your hand around the little toe area . . . make sure there are no wrinkles and then pull it back up. Check the heel area. We don't want any sign of a wrinkle about it . . . The wrinkle will be sure you get blisters, and those blisters are going to make you lose playing time, and if you're good enough, your loss of playing time might get the coach fired." After the sock was one with the foot so as to not cause blisters, Wooden would focus on the shoe, "There's always a danger of becoming untied when you are playing," he said. "If they become untied, I may have to take you out of the practice, I may have to take you out. Miss practice, you're going to miss playing time and not only that, it will irritate me a little too."

How many Division I college basketball coaches teach players to put on their socks and tie their shoes? Wooden did. He didn't want blisters, because blisters can slow a player down, lead to distractions. To that end, one of the first practices every season Wooden would lead his team through the proper technique for putting on socks and tying shoes. In safety sensitive environments, what level of detail do we need to have to prevent injury or incident? What is our equivalent of socks and shoes? I think if Wooden had been a safety professional, he would have said, nothing is too small to make a big deal out of . . .

Be Quick, but Don't Hurry—There is probably no more famous John Wooden quote than this one, "Be quick but don't hurry." This saying took on new meaning this winter as my son, age 8, played his first year of organized basketball. The game moves so quickly that early in the season when my son, and most other players, handled the ball, they were frozen. They couldn't decide if they should shoot, pass, dribble . . . they were thinking through each move, processing it in their heads. Near the end of the season however, players were more comfortable and could make decisions much quicker, many times without thinking. Our workers, and Wooden's basketball players, are at the top of their game. They are experienced, can anticipate the 'next move' and often move at a very fast pace. Wooden understood that when experience reaches a certain level, complacency can set in because workers, I mean players, act without thinking—Wooden called this 'hurry.' And Wooden knew that acting without thinking on the basketball court led to turnovers and poor shots, in safety it leads to near misses and injuries. There are no more famous quotes from Wooden, and there may be none other that applies so well to safety. Be quick; think before you act.

Integrity—"You can't live a perfect day without doing something for someone who will never be able to repay you." John Wooden. Wooden grew up in Indiana and was an All-American basketball player at Purdue, which is located in Indiana. He and his wife Nell loved the Midwest. For them, it seemed to 'fit' there style and personality. When Wooden applied to coach at UCLA, he also submitted an application at the University of Minnesota. He and Nell had discussed it and decided that, if offered, they preferred the Minnesota job. Minnesota officials called John, but due to a terrible snow storm the call didn't reach him, so he accepted the UCLA position. The next day, word reached John that he could take the Minnesota job but he said 'no thanks,' he had given his word to UCLA. Two years later, in 1950, the Purdue head coaching job opened, it was Wooden's 'dream job.' However, he was just starting the third year of a three-year contract and Wooden thought breaking

Operational Excellence
Sustained zero incidents and moving beyond zero
Vulnerability
Openly sharing personal experiences
Organizational Accountability
An attitude that asks, "what more can I do for safety."
Trust
Coaching/ Feedback
Support
Assume Positive Motives
Engagement Involvement
Appreciation
Safety Process
Safety Success Pyramid
Concept from John Wooden
Copyright 2010- -Kcrof Industries, LLC

the contract was breaking his word, he stayed at UCLA. In safety, our workers are always trying to find out what is motivating management. Is it money, production, corporate? Maybe it's integrity? And if it is, we are building a championship program. How will our workers know? They will know when we keep our word through very tough decisions . . . like Wooden. Leave your organization better than you found it . . . that stewardship and integrity can't be repaid.

The Process of Greatness—John Valley who played on the 1969 and 1970 UCLA Championship teams said, "On the first day of practice, I remember him saying, 'I'm not going to be talking to you about winning or losing because I think that's a by-product of our preparation. I would much rather be focused on the process of becoming the best team we're capable of becoming." How many years did Wooden coach UCLA before winning his first national championship? Fifteen! What were he and his coaching staff doing in these fifteen years? They were building the foundation for sustained results. In fact, in 1948, before Wooden accepted the head job at UCLA, he scribbled down what is now known as the Wooden Pyramid of Success. Atop the pyramid is 'competitive greatness.' On the levels leading up to this achievement are key milestones like loyalty, friendship and team spirit. Safety's end game is world-class safety performance or operational excellence. It means sustained greatness with zero injuries. To that end, I have included the Safety Success Pyramid. It's taken from Wooden's model and molded for safety. Use it, use Wooden's or make your own, regardless, begin today to build that foundation for long-term success.

8 What Day Will You Get Hurt?

It's About an Attitude, Not a Day . . .

Can we statistically determine what day our workers will get hurt? And, can the day of the week that an injury occurs mean that severity will be less, or greater? Over the last few years, there have been some industry experts who have predicted that injury severity increases on the day before a weekend or the day before an extended break. Can a quick internet search support this theory? Let's take a look. On Saturday, May 8th, a gas explosion in China's Hubei Province mine killed 10 and injured six. A weekend. A week later, China experienced another explosion and coal mine disaster, this time it happened late week, on a Thursday. In this case, 21 miners were killed. Do we have a late week trend? On further searching, we find that the BP gulf coast incident happened on a Tuesday. That seems to blow the late week theory. Or, the space shuttle Challenger exploded on a Tuesday as well . . . this myth might be busted.

In looking at a number of other incidents, injuries, and fatalities, we probably can't determine a statistical probability of an injury happening on a specific day of the week but if we look closer, we can find some themes. And, discovering these themes will allow us to be aware and prepared, ahead of a potential disaster.

End of a job—The single deadliest event on Mount Everest was on May 10, 1996 when seasoned and experienced guide Scott Fischer and seven other climbers were killed . . . they were on the decent. At the beginning of a hazardous job, we tend to be on our toes. We plan. We are vigilant. Our awareness is heightened. Yet, once we reach that peak, we tend to think the major hurdles are behind and we can let that guard down a little. Forbes associate editor Christopher Helman wrote the following in a recent article about the BP disaster, "We know with some certainty that workers were in the final stages of setting the final sections of pipe (production liner) in the hole and cementing it in place. The plan was to set cement plugs in the well, temporarily abandon it, and move the Deepwater Horizon off to a new drilling site within a couple days." Did the fact that they were 'coming down the mountain contribute to the incident? I'm not sure, but the end of a job or task can mean that we let our guard down allowing injury or disaster to creep in. Take extra precautions; both climbing and descending dangerous tasks.

Change of, or Absence in, Supervision—"Although it wasn't, May 2, 1972 almost felt like a Friday for the 173 miners reporting to their normal 7 a.m. to 3 p.m. day shift," Matt Forck writes in his new release, *Check Up From the Neck Up—101 Ways to Get Your Head in the Game of Life*. "The atmosphere at Sunshine Mine in Kellogg, Idaho, probably felt different because the top brass was several counties away attending the annual stockholder's meeting. With the 'big bosses' gone for the day, it seemed that everyone was taking it a step slower." It was early in the shift that the fire alarms rang. A fire in a mine can lead to disaster but this was a silver mine and didn't offer much fuel for a fire. Tom and Ron, partners for the last several years, left their post and headed to the man lift to go topside. On the walk there they joked that this might even mean an early beer at the local pub. Once at the man lift, waiting with dozens of other men, Ron collapsed; overcome by fumes. Tom grabbed him and pulled him back near their work location, to fresh air. Once Ron was feeling better, they again headed to the man lift. What they found there horrified them. All of the other men waiting for the lift, just minutes earlier were joking and laughing, were now dead.

Things change when the boss is out of town. When management shifts, a new boss is hired, one retires or someone is temporarily upgraded to fill a role. Attitudes change when management is off-site at an event, all day meeting or stockholders meeting. These situations can't be avoided but when they occur, be aware of job assignments, crew assignments and production rates. Instruct those leading the work to take extra time planning. It's even a good idea for the management left behind to be active in the field or on the floor, just to make sure work is progressing safely.

By the way, eight days and over 200 hours later, rescue crews reached Ron and Tom. Once safely above ground they learned they were the sole survivors of one of the worse mining incidents in the United States; an incident that took 91 lives.

A Simple and/or Routine Job in Combination with Weekend or Break—

It seemed to be an easygoing Thursday morning. It was in a safety committee meeting when my phone rang. I first ignored it, intending to dedicate my energies to the meeting. Yet, the phone rang again, and then again. I stepped out to take the call. It was the regional dispatcher. He told me that we had an electrical contact. He informed me that emergency services, including the life flight helicopter, were on site. I left the meeting and made the 80-minute drive to the location. I found that the crew was on their last day before a three-day break. I also found that the utility crew was working a very simple pole change-out job; one that each of the six men had done, dozens, if not hundreds of times. In the incident, two of the men had been electrocuted; one didn't make it.

While I may not be able to prove it statistically, I believe that there is something to the notion that opportunity for incident severity increases before a holiday or extended break, in this case a three-day break. But, I think that a combination of a break with a simple and routine task is the combination to watch out for. When this combination occurs, take some extra time in the job planning. Make sure the entire crew discusses all hazards and takes the appropriate actions to eliminate each hazard, according to the rules and policies. Finally, stop the work periodically to make sure everyone is still on the same page and rules are being followed.

End of the Day, End of Job Work Pressure—I was on cloud nine! I had just

finished my first presentation at a national safety conference. Since I wasn't flying out until the next morning, I walked the Baltimore Harbor waterfront, located a terrific seafood restaurant and was seated at a window, so I could watch the boats bouncing in the harbor. I was somewhere in the middle of my salad when the cell phone rang. It was a good friend, a safety professional with a utility, and he needed someone to talk to. He told me about a utility incident that happened just hours earlier. A service worker, at the end of his shift, was asked to install some labeling in a piece of energized electrical equipment. Although the job was very simple, the service worker apparently hurried to complete it. In the process, he contacted energized high voltage equipment. He was in a burn unit, clinging to life.

We feel pressure to hurry, to get it done. And this pressure is never greater than at the end of a shift. When we are racing to complete a job near shifts' end, take a few seconds to stop and perform a safety stop. A safety stop is when the entire crew stops, reviews the work and the safety work rules associated with the task, and then continues. This 90-second safety stop can literally save lives at the end of the day.

Remember, maybe the most famous end of shift work pressure incident was the Space Shuttle *Challenger* disaster. As you remember, the Challenger splintered into millions of pieces when it blew up 73 seconds after liftoff. To meet a pressure packed deadline, the decision was made to launch after some engineers questioned how an O-ring seal in its right solid rocket booster would respond in the cold weather. If there wasn't pressure of the 'deadline' would that decision have been different?

I'm not sure we can statistically prove that injuries or incidents will happen on specific days or at specific times, yet there are some warning signs to look out for. I think that there is a tendency to let one's guard down after the 'heavy lifting' on a job is finished and we are coming down the mountain. I think that when supervision shifts, we have simple tasks before a long break or we are hurrying to finish a job before the shift ends, all present certain dynamics that can lead to an incident. As safety leaders we need to be ready and aware that certain conditions make it easier for an injury or incident to occur. Where are those conditions in your work environment? And, what are we prepared to do to prevent 'bad' stuff from happening?

9 Showing Warts
How Living Above the Line can Lead to Safety Results!

In August of 1997, I completed my apprentice work and passed both the skills and written tests. I was a journey electrical lineman! About two months later, in late October, a huge windstorm hit the small Missouri town in which I lived. To no surprise, the phone rang. Lights were out. It was time to go to work.

I learned that we had several hundred people out of lights. I was given a handful of orders and began to work them, going to homes to check meters and services, clearing limbs and refusing cutouts. I was feeling pretty good about things. One rule during the apprenticeship is that an apprentice never works alone; there is always someone there to help and lend a watchful eye. Once topping out as a journeyman however, I could work alone and it was common during storms like this one, to do just that. Despite the fact that I was working alone for perhaps the first time, I was feeling good, empowered. In the middle of one order the dispatcher called. The 34.5 kV circuit that fed three small towns across the Missouri river had gone out. I needed to go to the substation and wait for further instructions.

By the time I arrived at the substation, evening had turned to night. The temperature had dropped another 15-degrees and the rain continued to pound. In talking to the dispatcher, we needed to open the breaker and associated switches so we could do some line repair across the river. Once complete, several thousand customers would be back in lights. In all, there were three specific switch orders, and I accomplished the first two without incident. The final order, to check open a certain switch, became a problem because I could not find the switch number. I looked everywhere. I began to hurry and became a little frantic. I knew the dispatcher was waiting for me, so was a crew across the river; their work could not begin until I finished these switch orders.

I was irritated and frustrated that I could not find the switch location. Finally, through the driving cold rain, I found it but there was no switch handle. I shined my flashlight through the heavy rain and substation steel looking for the position of the switchblades but with the weather being what it was, I couldn't see a thing. I was near frantic now, looking for a switch handle so I could operate the switch. Finally, after what seemed like forever, I found a handle, which is really a long metal tube. I inserted it into the switch housing and operated the switch. It was about half way through that something just felt wrong. As I finished the operation, I immediately knew that I had closed the switch, which was exactly what I was not supposed to do. Because I was hurrying and annoyed, I had just closed the bypass switch, reenergizing the line. Without thinking, I immediately threw the switch open.

Have you ever unplugged a toaster or an appliance and saw a small spark as you pulled the plug from the outlet? In the electrical business, pulling the plug from an outlet when the appliance is running is called dropping the load. When I opened the switch that I had just closed, I dropped the load. But this wasn't just a toaster; this was a couple thousand homes or several million toasters. You can imagine the spark, or in this case, electrical arch, that immediately unfolded right above my head. A huge fire ensued. The noise was intense. I ran!

"Vulnerability," Keith Ferazzi, business leader and author says, "is the courage to reveal your inner thoughts, warts and all, to another person." I think the interesting thing about this story, in addition to the fact that I did not get fired, is that I never shared it. A couple of years after it happened I was promoted to a line-manager. And, a few years after that I moved into the utility's safety staff. In this story there are some great lessons for others to learn from, but I didn't share because I was afraid to be vulnerable—I was afraid to live above the line.

If we, as safety leaders, safety professionals and managers are going to be successful, we have to learn to share our experiences, be vulnerable and stay above the line. Below are four quick pointers to living above the line.

It's About Relationships—"Rules without relationships," said Josh McDowell, writer and speaker on family values, "creates rebellion." One of the best ways to build relationships is to show that you too are vulnerable. In sharing personal experiences over time you will build the trust and credibility for when safety rules and work practices need to be challenged or changed.

Shadow of the Leader, Share Near Miss Events—According to some estimates, near miss events may cost twice as much as serious incidents or fatalities. According to a Houston Business Bureau, CII and Exxon Chemical report, a near miss event is estimated to cost about $1,300 and they estimate about 1,000 near miss events for every fatality. Using 2004 Bureau of Labor Statistics data, 5,703 workplace fatalities were reported across the United States. At an estimated million dollars per fatality, near misses cost the private sector more than a trillion dollars more, on the monetary side of the equation, than fatality. Each of us have previous work experiences or have simple experiences from a weekend of yard work, installing Holiday lights or painting that we can bring back into work on Monday and share. Our organization will only cast a shadow for sharing near miss reports and vulnerability if we do it first.

Strive to be Above the Line—Remember the chart on page 8? The same concept applies here. That chart does a great job in helping us understand just where the line is in terms of vulnerability. We must constantly strive to lead above the line. To do that, we must think about sharing and being transparent, instead of withholding. We must coach, appreciate and encourage instead of tell. A story can go a long way toward compliance instead of simply stating the rule. We have to lead with the spirit of giving and volunteering instead of taking and/ or withdrawing.

Constantly Measure—The 'Above the Line' chart not only allows us to benchmark specific actions that lean toward vulnerability, it helps us measure our choices throughout the day against 'the line.' As leaders, safety professionals and managers, we have challenges from all directions. The chart allows us to reflect after a meeting or critical conversation and specifically measure our ability to work above the line in that situation. The more consistently you and your team can work above the line, the quicker results will follow!

"Laugh at yourself, but don't ever aim your doubt at yourself," Alan Alda once said. "Be bold. When you embark for strange places, don't leave any of yourself safely on shore. Have the nerve to go into unexplored territory." What stories, events and experiences have you been holding onto, unwilling to share? Whatever it is, it can't be much worse than setting fire to a substation. Share today. Be vulnerable. Make a commitment to live *above the line!*

10 How to Launch!
Safety Lessons Learned from Apollo 11

"That's one small step for man," Neil Armstrong said from the lunar surface on July 20, 1969, "One giant leap for mankind," Of course I studied this phrase in school but only now, some four decades removed from Apollo 11's mission to the moon, begin to fully understand just how big of an accomplishment this was and still is. Norm Mailer, being overwhelmed by the experience later wrote that Apollo 11 was, "Mankind having found a way to talk to God." Arthur C. Clark who wrote the book, *2001, A Space Odyssey* said, "At liftoff, I cried for the first time in 20 years and prayed for the first time in 40 years."

My generation, (I was born three months after the landing), along with those generations after mine, find it hard to fully understand the accomplishments that Apollo 11 represents. Yet, in a time without personal computers, cell phones, iPods, satellite television, GPS or microwave ovens, our country sent three men to space, landing two of them on the moon, then returned them safely. What happened, and how it happened, still has great value to today's search to launch into excellence, continuous improvement and safety results. Let me explain.

Launch Through Engineer—Several years ago I was in a meeting with fellow safety professionals. The discussion centered around noise monitoring data. It was determined that noise levels exceeded acceptable limits. The conclusion was workers should be furnished hearing protection. It was all but decided until someone spoke up and asked, have we discussed engineering out the hazard? The room was silent.

The principles of safety are easy, engineer, educate, enforce. Yet today, many leaders, safety professionals and companies quickly move away from engineering toward education and enforcement. Just like in our discussion of hearing protection, PPE is all too often the 'quick fix' that allows the hazard to exist, waiting to hurt someone one day while engineering solutions can eliminate the hazard completely. Performance aerodynamics engineering and development director for Apollo 11 said this, "If I am an engineer I better dam well understand what reliability and what failure means, otherwise I am not an engineer. We had redundant valves, quad redundant valves everything else. I basically said the best way to deal with risk management is in the basic conceptual design, get the dam risk out of it. And I think that is what made the program a success." "Science is about what is," Neil Armstrong once said, "and engineering is about what can be" What 'can be' in terms of engineering solutions in your workplace?

Questions to Launch—Where have you and your team quietly slipped into PPE mode? Do you need a team to look at engineering improvements? What engineering opportunities exist today that can make your work environment safer?

Launch By Training—In your next safety meeting, sit in the back and observe. After several minutes ask yourself if you are at a funeral or a safety meeting. Question if you are with someone in the waiting room, about to get a root canal, or a training session. The fact is that many of our workers have been around the block a time or two. They are experienced and seasoned. And, with this also comes complacency in training and learning. In most cases, our people have worked safely so many years that we lose the connection between something really bad happening on the job and our training to prevent that occurrence. This means that training all too often leads to a 'check the box' attitude.

"The training regiment, however, was exhausting on every level of human effort, for both astronauts and their ground controllers," writes Craig Nelson in his epic book called *Rocket Men: The Epic Story of the First Men on the Moon*. "After learning the basics on a sim-flight that went smoothly, both sides were subsequently put through their paces with a series of problems like mechanical breakdown, conflicting data streams and all out system failures." Nelson writes later in the book about simulation testing, "This process worked so well, in fact, that in time many astronauts would calm themselves in real work crisis by thinking, *this is just like simulation*." Because of their training, they had great confidence!

Questions to Launch—What confidence do we have that our training is pushing our workers to be better? Are we using simulations and activities or simply sitting and listening? How can we shift to a continuous learning environment keeping that close connection between the need to train and hazard elimination?

Launch Through Toughness—It is no secret that feedback is the 'lifeblood' of continuous improvement. NASA, like any organization, went through periods when feedback did not occur. When that happens, tragedy often results. It happened with Apollo 1 when three astronauts were killed in a test simulation. Later, a controller said:

"I tend to be maybe one of the more emotional of the controllers. I believe that is part of a leader's responsibilities, to get his people pumped up. I gave what the controller's came to know as the tough and competent speech. And concluded the talk by identifying the problem throughout all of our preparation for Apollo 1 was the fact that we were not tough enough. We were avoiding our responsibilities. We had not assumed the accountability we should have for what was going on during that day's test. We had the opportunity to call it off, to say this isn't right and shut it down and none of us did. We had become very complacent about working on a pure oxygen environment. We well know this was dangerous. Many of us who flew aircraft knew it was extremely dangerous but we sort of stopped learning. We had just really taken it for granted that this was the environment, and since we had flown the Mercury and Gemini program at this 100% Oxygen environment everything was okay, and it wasn't. I had each member of the control team on the black boards in their office write tough and competent at the top of the blackboard and that could never be erased until we had gotten a man on the moon."

Questions to Launch—Feedback, a gauge of being 'tough and competent.' How tough is your organization? Where have we stopped being responsible and accepted the 'status quo' in dangerous environments? How tough and competent is your culture?

Launch Through Greatness—Neil Armstrong said this after the Apollo mission:

"I was certainly aware that this was a culmination of the work of 300,000 or 400,000 people over a decade, and that the nation's hope and outward appearance largely rested on how the results came out. With those pressures it seems the most important thing to do was focus on our job as best we were able to and try to allow nothing to distract us from doing the very best job we could. And you know, I have no complaints the way my colleagues were able to step up to that.

Each of the components of our hardware were designed to certain reliability specifications. And for the majority, to my recollection, had a reliability requirement of 0.99996, which means you have 4 failures in 100,000 operations. I've been told that if every component met its reliability specifications precisely, that a typical Apollo fight would have about 1,000 separate identifiable failures. In fact we have like 150 failures per flight, better than statistical methods would tell you that you might have. I can only attribute that to the fact that every guy in the project, every guy at the bench building something, every assembler, every inspector, every guy who was setting up the test and cranking the torque wrench and so on is saying, man or women, if anything goes wrong here, it's not going to be my fault, because my part is going to be better than I have to make it. And when you have hundreds of thousands of people all doing their job a little bit better than they have to, you get an improvement in performance. And that is the only reason we could have pulled this whole thing off.

The way that happens, the way that made it different from other sectors of the government in which some people are sometimes properly critical is that this was a project in which everybody involved was one interested, two dedicated and three fascinated by the job they were doing. And whenever you have those ingredients whether it be government or private industry or retail store, you're going to win!"

Questions to Launch—The fact is that you wouldn't be in business unless you were making something or doing something that someone needs for something; in short, you and what you are doing are extremely valuable. How do you capitalize on the value you bring to your market and your community? What can your team rally around? How can we foster that 'my part is going to be better than I have to make it' attitude?

Launch Through Deadlines—There was great power in President John F. Kennedy's statement in May of 1961 when he said, "I believe that this nation should commit itself to achieving the goal, before the decade is out, of landing a man on the moon and returning him safely to earth." In fact, Milton Servero, Aerospace engineer recalls, "I always noticed that when President Kennedy said 'go to the moon by the end of the decade' that all of our badges issued to us expired on December 31, 1969. Well, that's a message for you, either do it or you are not employed anymore."

It's one thing to set a deadline and another to achieve one's goal. So how did NASA achieve such an accomplishment? Well, they had support, teamwork, and engagement toward a common goal . . . they were aligned.

Questions to Launch—What is your goal around safety, and is it 'big' enough to be inspiring? Is there support around your safety goal? Is everyone in your organization aligned around that goal?

"That we could step beyond our narrow personal concerns to achieve great things," wrote Bob Herbert after the Apollo 11 mission, "That we could do better, be better, if only we had the strength and courage to work harder and dream bigger." There are great lessons from Apollo 11's accomplishments some four decades ago. Lessons learned on so many fronts including engineering, training, toughness, greatness and deadlines. Can we use their lessons go be great? Yes, we can . . . if we too want to launch!

11 The Safety Communication Awareness Tool (SCAT)

How to Reach the Top of Safety Communication

A number of years ago, I was asked to be the opening day honorary coach for a little league baseball team. For the opening game, I would be their motivational leader; I would also have the honor of being third base coach and have the responsibility for drinks and snacks.

In this particular small town, little league baseball is a big deal. On opening day, players from the youngest age group, six and seven, all the way up to the 13 and 14-year-olds parade through Main Street. Riding on trucks and trailers. These players throw candy and wave to family, friends and on-lookers. After the parade, I met the team. Gathering them in a huddle, I introduced myself. "Today, if you are batting and someone is on base, you look at me," I instructed the boys in a stern voice. "I will give you a series of baseball signs. Based on this sign, you will know if you should bunt, try to walk or swing away." These 9 and 10-year-old boys were looking at me like I was nuts. After a long and series pause, I smiled and said, "I'm kidding boys, we are here to have fun. If you want to walk, you walk. If you want to try for a home run, swing away." With a long sigh of relief, we all stacked hands and yelled 'team!'

Now, this has probably never happened in this small town's little league baseball history, but due to my positive pre-game pep talk, we retired the other team in order. When it was our turn to bat, our first hitter took a walk. In my short time coaching, Josh, this next batter, seemed to be a bright and athletic kid. As he stepped in the batter's box, I flashed some pretend signs, just for fun. Josh played along as he watched my signs and nodded understanding . . . the other team didn't know what to think. Yet, Josh was about to throw the biggest curve ball of all. Right before his first pitch, he yelled 'time out.' The umpire yelled 'time,' and Josh came jogging toward me.

Now I'm thinking, Josh, what are you doing, they are just pretend signs but I didn't panic. As Josh walked over, I got down on his level and said, "Josh, my friend, what are you doing? They are just pretend signs." To which Josh responds with a smile, "Yeah, I know Coach Forck, I'm just pretending like I didn't read them!"

Think about it, line managers, foremen, superintendents and managers are all responsible to 'get the job done.' In so doing, they send signals, dozens an hour. These signals tell their employees what is important and what is not. These actions tell their people what needs to be done and what can be skipped or modified. More times than not, these decisions, actions and signals involve safety sensitive tasks. Safety is a game of signals and signs. Our people don't read our lips but watch our feet—in other words, it is not what we say but the sign and signal we send. And, the SCAT (Safety Communication Awareness Tool) model can serve as a tool to measure your communication.

A Typical Day . . .

The SCAT model can be called a ladder or elevator because everyone sends dozens and dozens of signals up and down the model during the course of an hour and day. For example, a line manager may lead a safety meeting where he is very prepared and engaged. From there he may go on the floor; intentionally going to one area of the floor, avoiding another because he is just a little uneasy about what he may find in that other area. Next, he goes to observe a new process, one that just started. He is unknowing about the safety procedures because he has yet to find time to read the job safety analysis; instead letting another supervisor take the lead. As he is walking back to the office, he notices an employee working at risk. He has talked to this employee before and the employee is about to retire, so he decides to wait and hope the problem will be

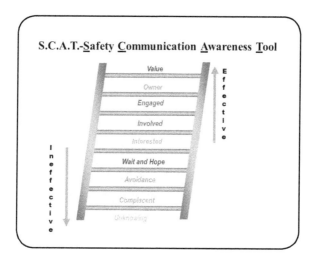

gone in a month or two. Walking on, he goes to new employees and pulls them from the line to instruct them on ergonomic principles. Is this typical? Probably more than we care to admit!

Using the SCAT Model

The SCAT model can give us great insight as an awareness tool:

Be Aware—The first key in using the SCAT model is to be aware that safety is a game of signals and our employees pick up each and every signal we send. Even passive signals or signals not intended for one employee will be quickly picked up, interpreted and passed on.

Be Interested—You may notice that 'interested' on the SCAT model is the mid-point and the beginning of effective. During busy and stressful times, it might be hard to master 'engaged' or 'owner' or 'value' but strive for 'interested'. It is a positive signal that does not subtract from your safety efforts.

Be Consistently Effective—The truth is that one or two signals in the ineffective range will negate dozens from the 'interested' range. To that end, we must strive to be consistently at interested and above. Consistency, over time, is the only way to achieve sustained results.

Be Accountable—Accountability can be defined as 'what more can I do to get results.' Now that you and your team have an understanding of the power of ones signals on safety performance and a model for which to base your communication, hold yourself accountable to the SCAT model. Partner with another person with similar responsibility so you can share struggles, provide coaching and give each other encouragement.

"Integrity," one could say, "is when what you say matches what you do." Safety is a game of signals. Often, we can be complacent, avoidant and even wait and hope. When we send these signals, and similar ones too, our safety efforts will be ineffective. Integrate—when it comes to safety, let what you say match the signals you send . . . and the SCAT model can help you do it!

12 Chicken Rings
What's in Your Circle of Safety?

Have you ever heard of a chicken ring? Even though I have earned a journey lineman card in distribution line work and like to run excavators, chain saws and skid loaders, I'm still a city slicker. So, when I was given my first chicken ring, I didn't have any idea what it was. It turns out chicken rings have at least two primary purposes. These small plastic rings that are about the circumference of a quarter, are used for chickens. The rings can be quickly slipped on and off of a chicken leg or neck to mark it. Apparently if you are a chicken and earn a chicken ring, it's a good thing—you get to live another day. The second use of a chicken ring, interestingly enough, is for safety.

My first chicken ring was given to me by Tom. Tom was a legendary electric foreman. His demands were high, he motivated his men and he had a knack for safety awareness. He purchased green chicken rings by the bushel. As Tom would conduct a job briefing or safety talk, he would insist that his men take a chicken ring or two. These rings would we slipped on hard hats, boots, door handles, steering wheels, bucket controls, etc. The point of the ring was to raise safety awareness—awareness to the Circle of Safety.

This simple green chicken ring is a great reminder of a simple yet powerful habit, a circle of safety. In the utility and construction businesses and in any safety sensitive environment, there are several times that we are taught about the 'circle of safety.' In our defensive driving classes, we are taught that our large trucks have a circle of safety. It's the zone around our vehicles where smaller cars can sneak into a blind spot. We are reminded to constantly monitor these blind spots to prevent an incident. After a job is finished, yet before one hops in the truck to drive to the next job, we are asked to perform a circle of safety. This is a complete walk around of the work site and vehicles to make sure our load is secure, bin doors closed and all tools are picked up and properly stored. We need to perform a circle of safety before we back our vehicles, since we can't see what's behind us. The point is simple; a circle of safety is an effective safety tool. A high level of safety awareness is needed to keep one focused on practices like the circle of safety, and a chicken ring is an effective reminder. To that end, let's take a quick look at the five key things to think about regarding chicken rings and circles of safety.

Five Keys to an Effective Circle of Safety

Remember Changing Conditions—Joe was in a hurry and he need to pick up some materials for the next job. He buzzed into the works headquarters, parked his bucket truck and dashed inside. He quickly returned with a handful of material, which he put away in the driver's side bin. He thought about a circle of safety but reasoned that he had been there less than five minutes, so he climbed into the cab and backed out—backing directly into another truck that had parked behind him, right in his blind spot. This story is completely true and hundreds of incidents like this happen every day in our industry. Remember, conditions change by the second and a simple walk around, or circle of safety, is the best way to prevent these events.

Look for and Eliminate Hazards—Everyone who has worked for any time in a hazardous field has a tragic story. As a former safety professional for a utility, I have a number of stories, but the one that stands out is when a crew inadvertently raised a boom into a 12kV phase. When the phase made contact, a worker on the ground was getting material off the truck and he received a fatal electric shock. Between the linemen on that job, there was probably over 100 years of experience. Everyone on the job saw the overhead wires. Everyone saw where the truck was parked. The only thing on the job that could have caused serious injury or death was the energy source—it was recognized but not eliminated with rubber cover. The circle of safety encourages that we not only identify hazards, which is the first key step to safety, but it requires the next step, to eliminate hazards. Through this circle of safety, maybe the next generation of utility workers won't have tragic stories to share like our generation does.

It's Not Just for Trucks and Poles—For years the utility industry has promoted a circle of safety when backing a utility truck and before climbing a pole, but this concept isn't just for trucks and poles. In fact, the principles of a circle of safety are easy; before you begin a task, walk through the work area, identify and eliminate hazards, then proceed. If we can use this easy, ' before you begin a task, walk through the work area, identify and eliminate hazards,' formula on each job and each task, we have a formula for safety success and injury prevention that can't be beat!

Expect the Unexpected—A number of years ago, an old troubleman completed a job in the back yard of a home. He then walked from the work location to the front yard and to the street, where his truck was parked. He knew he would need the same tools for the next job, so he set his tool in the passenger floor board. He did not walk around the truck, instead, jumped in and drove to the next job. As he got out of his truck, he heard crying—it was coming from the back of his truck! He ran to the rear of the truck to find a young boy lying there in tears.

It turns out that while the troubleman was in the back yard working, the young boy was playing on the truck. When the troubleman returned, the boy panicked and froze, remaining on the truck. The incident could have been tragic if the child had chosen to jump off the moving truck. And, the whole incident could have been avoided by a simple walk around.

Chicken Ring It—One of the keys to performing a circle of safety is aware-ness. Utility workers need constant awareness, and one of the best ways to create awareness is through a visual reminder—this brings us back to the chicken rings. After 40 years in the utility business, Tom is retiring this year. It is a running joke that his biggest accomplishment over these four decades isn't in the utility world, instead, he single-handedly kept the chicken ring company in business by purchasing thousands of these rings. I'm not sure if that is true, what I do know is that he had a great impact in the safety of his workers by teaching the circle of safety and handing out chicken rings as a visual reminder. It's not about Tom, or the chicken ring. It's about keeping our people safe through a simple yet memorable awareness activity, so that each of our people can leave work as healthy as they arrived. What is your chicken ring?

13 Far Forward, Safety
Where Leaders Dare to Go

Dr. Jadick was anxious. It was the evening of November 8th, 2004 and Jadick was the Battalion Surgeon for the 1st Battalion 8th Marine regiment. His Marines, along with other Marines and military personnel would be moving against the strongest insurgence of the war, those holding out in Fallujah. In mere hours he would be treating wounded, seriously wounded. Jadick knew this would probably be the most brutal urban fighting Marines had experienced since Vietnam. As with any medical professional about to be presented with major traumas, Jadick was anxious about his own ability. Would he be able to handle each case, what about his pace of work if there are multiple seriously wounded? How would his Corpsmen do, and did he prepare them well enough? Yet, there was an overarching concern that Jadick had; he worried that his aid station was too far in the rear. It was a 45 minute ambulance ride away. In addition to his other fears, he really thought this time gap, between battle-field wound and professional care, 45 minutes, was too great. The answer to his questions was just hours away.

"Traditional battlefield medicine," Jadick wrote in his book entitled, *On Call in Hell*, "Was developed over hundreds of years and it made sense for earlier times. The line between KIA (Killed In Action) and WIA (Wounded In Action) was more absolute then. Wounds fell more easily into the categories of lethal or nonlethal. And corpsman or medics with the units served as much to provide comfort to the dying and transport to the dead as to intervene with lifesaving care. Both of those rolls are still an essential part of the job, but today we have more wounds that can go either way. And we have vastly improved higher level care that can help severely injured warriors survive and rehabilitate off the battle field. The interesting thing is that the technology and techniques of battlefield medicine have not changed much since Vietnam."

Jadick's team received a call early the morning of the 9th, "Marine down." Jadick didn't stay in the rear, instead he moved with the ambulance team to help give care to the causality. They pulled up to the injured—it was a bad wound. Other calls came in and the ambulance team was needed to aid others. In the end, Jadick turned away one less seriously injured Marine, because the ambulance was beyond capacity. The first Marine, who was very seriously injured, didn't make it, dying on the ride back to the aid station. Jadick's fears were confirmed. He and his aid station were too far back—he would move them closer.

It took permission, but once Jadick proved that he and his select corpsman could travel light, they were allowed to move forward and set up a forward aid station—literally in the middle of the fight. Later, Lt. Col. Mark Winn estimated that without Jadick at the front, the Marines would have lost an additional 30 men. Of the hundreds of men treated by Jadick and his team at the newly established forward aid station, only one died after reaching a hospital. Overall, 53 Marines and Navy SEALs died in the battle. What Jadick led, and what he and his corpsmen did was truly heroic. Not only did they save lives, they set a model that will surely be used moving forward. And, in Jadick's leadership there are a few lessons for us safety professionals, as well.

Improve Your Position—Jadick was constantly working to improve his position—in order to save lives. At the forward aid station, he and his corpsman worked to position sand bags over windows and openings, to protect against sniper fire. They worked to set up latrines. They practiced routines that moved needed medical supplies with great speed and found concrete blocks that could be used as bases for operating tables. In short, they continually looked to improve their position. What do we do to improve our position, to improve safety of our people? We often are caught in the trap of looking for the next big thing instead or major process improvement. I think Jadick would encourage us to pursue them, just as he moved the aid station to the front lines, but he would also suggest that each day we should lead some small change that improves our people's position (safety).

Define the Golden Hour—Jadick wrote, "In treating traumatic injuries, there is something known as the golden hour. A badly injured person who gets to the hospital within an hour is much more likely to be saved." But Jadick knew that in combat the "golden hour" doesn't exist. Left unaided," said Jadick, "the wounded could die in 15 minutes, and there are some things that could kill them in six minutes. If they had an arterial bleed, it could be three minutes." To that end, medicine's 'golden hour' has changed – it is mere minutes to save a life. What is safety's 'golden hour'? What is the time between a near miss and an actual injury if the hazard isn't eliminated? How much time exists between a defective piece of equipment being identified and it causing an injury, if not taken out of service? Or, what is the time gap between an unsafe act and injury? Jadick understood how crucial it was to establish the right care in the right time—what is our 'golden hour.'

All Leaders to the Front, Please—After Fallujah, Jadick wrote, "There were real dangers and that raises the issue of whether by going in to establish the FAS (Forward Aid Station) myself that I put an important battalion asset, me, the doctor, needlessly at risk. I think that is looking at it all backwards. By putting me forward I thought we were maximizing our assets. First there was the leadership issue. A battalion surgeon isn't just the doctor, he's also the leading officer of a medical platoon and no self-respecting platoon leader, anywhere, is going to send his men in to do a job he wouldn't do himself. There is a big difference between training and experience." Are our leaders at the front? Do they make a certain number of field observations each month, week or day? Is it required as part of their performance appraisal? If we are going to win the safety battle we need our leaders to establish safety's equivalent of a FAS. A forward station staffed with leaders who quickly recognize hazards and at risk actions and take corrective action—no questions asked.

What is Your Sphere of Influence—"I couldn't control who got hit and where but I still had my sphere of influence, and I decided that if it was taking too long to get the wounded out of the city then the only way we could cut the travel time down was to move ourselves in. That would mean in effect setting up an emergency room in the middle of the hot zone," Jadick reflected. In the workplace, we can't control who makes a poor choice on a given day, nor what equipment may fail at any one point. But, we can, and must, exercise our 'sphere of influence' just as Jadick did. As he moved his care closer to his men, we must move our care, coaching and knowledge even closer to our front line workers. If that means setting up our safety office on the plant floor—I believe that Jadick would approve.

In the end, Dr. Richard Jadick was honored with a Bronze medal with Combat V for Valor. The Valor distinction denotes "Those individuals who were awarded a decoration in recognition of a valorous act performed during direct combat with an enemy force. It may also denote an accomplishment of a heroic nature in direct support of operations against an enemy force" Later, Winn said of Jadick, "I have never seen a doctor display the kind of courage and bravery that Rich did during Fallujah." Jadick shakes off the praise and instead puts it all in the category of 'just doing my job'. But in this 'just doing my job' attitude Jadick saved dozens of lives. And, in the process teaches valuable lessons to each of us. Thank you Dr. Jadick!

14 How are Your Executive Safety Skills?

My daughter has moderate to severe Attention Deficient Disorder (ADD) and Sensory Integration Disorder. My wife and I saw signs early—we weren't sure what these signs were but we saw signs. Because of these 'signs,' from the time she was a toddler, she is ten years old now, we built a team around her and around us. This team included doctors and professionals of all types who could give us advice and guidance so that we could help her succeed.

On a recent visit to one of these professionals we were discussing home work. Out daughter's third grade teacher expects all of the students to copy tasks and homework assignments into a day planner. If the assignment is completed throughout the day, then it is to be 'x-ed' out in the planner. If it is not completed, then it is a homework assignment and will need to be completed that evening. Stephanie and I were explaining to the professional that our daughter didn't seem to have the organizational skills to consistently copy the assignment from the board, then mark it off her day planner if it was completed, or if not bring home the books and papers needed for the homework. It sounds simple to us as adults but she was struggling with the concept.

The professional, who was listening carefully, finally smiled and said, "Don't worry! Asking any third grader, let alone one who has added challenges of ADD, to do what is being asked is very difficult. Most third graders do not have the executive skills to carry out this task." He explained,

"This is something that I would see develop consistently in a fifth or sixth grader."

Since this visit, we have used the term 'executive skills' a lot. It helps us understand that our daughter, and all children; develop key skills over time, each at their own pace. The concept of 'executive skills' however is not limited to the development of children. It applies to all sorts of development, especially utility work and safety. Our workers are all at different stages of development and in different stages in life. We have new employees and apprentice workers, skilled journeyman and wise foreman and crew leaders. Each stage is still in development and each stage is still working on his or her executive safety skills. The following are some key executive skills to keep us safe!

Learning From the Mistakes of Others—A rule of the trade is to 'never make the same mistake twice.' Yet, we spend a lot of time in safety meetings reviewing near miss and injury reports—are we learning from the mistakes of others? How often do we change our behavior because of the mistakes of others? Learning and changing because of what others have experienced is a key executive safety skill!

Speaking Out—Recently I took a friend to pick up his car at a dealership. His car had been in the garage and was now ready. I was a decent acquaintance of the owner of the dealership, so I went in to see if he was around. The walk from my car to the owner's office was very short but on that walk no less than three people stopped me to ask if I needed help or assistance. One of these people seemed to be the maintenance supervisor, not related to sales. The owner wasn't there, but the next time I saw him I told him that his customer service is outstanding—relaying the experience of his team being so helpful. He said that years ago he had implemented the 10-foot rule. That means no matter who you are, from the top sales person to the janitor, if a customer is within ten feet, you stop and ask if they need help. And, from the perspective of a recent customer, it is impressive!

In the last couple of years, so much has been made in the safety arena about coaching and feedback or simply speaking out when it comes to safety. I have learned that this is an executive skill that everyone can learn. Whether it is speaking out to a customer, in the case of the car dealership, a co-worker regarding safety, it is an executive skill that makes a noticeable difference.

Brake—Having ADD is creative energy all of the time, a fun light-hearted spirit and intense focus on one thing now and another a split second later. ADD kids are bright and fun, but one executive skill that is a must is the concept of a brake. When our daughter is 'fast' or off track, we encourage her to 'brake'. That allows her to stop and refocus on what is most important at the time. Using ones brake when gloving primary, flagging traffic or performing other safety sensitive jobs is very important. It allows us to stop, if only for a few seconds, check all of the hazards and make sure we are getting it done the right way. After you have braked and checked your surroundings, you can safely continue. Braking is an executive skill that must be learned. Stopping a job or task to reevaluating the hazards and any changing conditions is a key to safety, and a key skill to master.

Thorough Planning—When I was an apprentice electrical line worker in Kirksville, Missouri I would be assigned from crew to crew. I recall one old line foreman who would always take the time to carefully review the job before anyone started work. As the new guy, they gave me time to ask questions, and I always had dozens of questions. As we ended the job briefing and went to work he would always say, "And remember, nobody gets hurt today!" Years later, as a safety professional for a utility, we found that failure to properly plan was a leading cause or a contributing factor to each injury. And the more severe the incident the more the lack of job planning played a significant role in contributing to the incident.

The point is, planning is an executive skill that must be mastered. Many utilities, construction companies, and contractors are now using job planning forms to make this more effective. No matter how you get results, job planning is an executive skill that you and your organization must have.

We're Getting Older—In our nearly twenty years of marriage, my wife and I have moved, on average, once every other year. We recently decided to do it again. Last summer we purchased a lot in a subdivision, and several weeks ago, we sold our house. The plan is that we are moving into a rental house, then we will start construction on our home—these things always sound better when planning than in practice! On moving day, we had a number of friends and several hand trucks (dollies) and we went to work. About half way through, someone stopped and said, "Man, my back hurts, we should have all stopped to stretch."

We are all getting older, this is a key point to remember for those who earn a living with their hands and the sweat of their brow. Getting older means buying into a stretch and flex program, whether offered by your employer or done on your own. It means eating right and getting the right amount of rest. What we could get by with in our twenties can leave lasting pain in our forties—if you don't believe me, just ask a few of the weekend warriors who helped me move.

You're Not Bullet Proof—I still remember my first call out. I had just topped out as a journey man and we were getting a strong October wind storm. We had lights out everywhere. I was feeling great, bullet proof, until I was called to the substation and operated the wrong switch causing a very large 34.5 kV fire! In time, all of us young journeyman learn that we aren't bullet proof. In truth, however, understanding the fact that we are not bullet proof today, instead of after an incident, is an executive skill worth learning. While no one was hurt by my switching error, it caused some additional work for us that night and some additional 'trouble' for me. What are you doing to rid yourself of the bullet proof attitude?

At the time of this writing, our daughter is just a few weeks away from finishing the fourth grade. She has had a great year. And, with the help of her mom (and sometimes dad) and the larger team, she is doing very well in school. Yet with each new year, there are new executive skills to learn in order to continue success. What are your executive safety skills? And, what new ones will you need to develop for continued success?

15 Fix the Vending Machine
How C.A.R.E.ing Leads to Results

Tom and I have been close friends since he and I worked together as safety supervisors for a major mid-western utility. Together, we worked storms, conducted training and bounced ideas off of one another. Tom recently shared this story with a group. I thought it was so good that I asked him if I could share it hear, and he gave me permission.

About 15 years ago, I was the Safety and Risk Manager of a large manufacturing company. One of the employees, an older and reserve man, told me the vending machine was broke and asked if I could contact the vending company to come out and repair it. I said no problem. I had every intention to contact the vending machine repairman, but other issues came up and I put the employee's request on the back burner until ultimately I forgot it. The following week the same employee came and indicated the vending machine still wasn't fixed, and it would be appreciated if I would contact the vending machine company to get them out to fix the problem. Again, other issues came up and I placed the employee's request on the back burner until I forgot it. Another week went by and the employee, very upset came to me again about the vending machine not being fixed.

I was put back by the fact that a simple machine with candy bars would cause that kind of outburst. But I quickly found out that it wasn't a simple machine that held candy bars, but a turnstile machine that provided sandwiches and chips. I learned that this employee's wife of 25 years had made his lunch every day. This employee would bring the same type of lunch that his wife made, every day for 25 years. Two years ago, the employee lost his wife after a struggle with cancer. So for the past two years he would buy his lunch from this same vending machine.

Leadership guru and author John Maxwell writes, "Your people don't care what you know until they know you care." It's easy to respond to the seemingly big stuff, the items that can cause great damage or shut the plant down. It's much harder to 'fix the vending machines' day in and day out. Yet, it is in these seemingly small situations that our people are watching very closely. Sure they are watching to see if we will get the job done. But more importantly, they are simply watching to see if we care.

Today, remember a simple model that can help you 'fix the vending machines' and C.A.R.E.

C: *Close the Loop*—As a line manager for a major electric utility I learned that I can't ignore the small stuff. As a supervisor, we take care of the big stuff right away. And, if it's big enough, we get a team to help us. But, if it is seemingly small, like fixing the vending machine, we often just don't get around to it. It is these seemingly small things that I ignored, forgot or let slide that would come back to get me, every time.

To that end, we need to close the loop for all of our employees. We need to set up a proactive system to ensure that things get done, especially the seemingly small stuff. If we can do this, we will earn more trust and confidence than we can find with any other action. Being good at the small stuff means that we care. And, if we can close the loop for our employees, they will close the loop for use, day in and day out.

A: *Assume Innocence*—Several months ago a good friend of mine called, he wanted to complain. He said that he had worked for several weeks to complete a project. Once complete he had an opportunity to present it to the senior leadership team for his company. He was calling because everyone liked it except one Vice President. This VP was critical of a couple of sections within the report. My friend commented, "This guy (referring to the VP) is just trying to make a name for himself, he'll do anything to get ahead!"

My friend and I go way back, so we can be brutally honest with each other, so I replied, "Yeah, maybe he is trying to get ahead, or may from his perspective, there is really an issue with these portions of your report."

When Tom's employee came to ask about fixing the vending machine, Tom probably thought it was a 'jab' at management, that things aren't working around the plant. On the second request, Tom may have thought that the request to fix the vending machine was personal, since he didn't get it fixed the first time. In truth, Tom's employee just wanted a lunch! We have to assume innocence with all of our employees and co-workers. We don't know what they are thinking, but chances are, they are honest thoughts, just wanting to improve or help out. About that VP, he may have wanted to get a head but he may have also wanted to make the report better, and my friend missed that opportunity, because he assumed something other than innocence.

R: Relationships—Take a moment to work through this simple five question quiz. First, for each of your direct reports, how many have children and what are their children's names? For your direct reports, who is married and what is their spouse's name? Do you know the primary interest or hobby of each of your direct reports? In the last 30 days have you given direct feedback about work performance and/or safety to each of your direct reports? Finally, in the last 90 days, have you done something 'special' for each of your direct reports, like a note on their birthday, a thank you card for a job well done, etc.?

Josh McDowell, a counselor and author of several books in building family writes, "Rules without relationships cause rebellion." If you scored a four or five on this quiz, you have probably taken the time to form a relationship with your people. If you scored a three, you are on the tipping point of forming a relationship or just being lucky. If you scored between zero and two you are working from rules and not relationships. Sustainable results come from relationship driven teams . . . what is your score?

E: *Engage in Coaching and Feedback*—"You can pretend to care," reads one of my favorite quotes from an anonymous author, "but you can't pretend to be there." Bruce was a minister and researcher. He approached a charitable board foundation a number of years ago. This foundation had a reputation for awarding grants for research on human effectiveness. Bruce's proposal was that he would tour the United States and Europe and interview the most successful business people and politicians of the time; asking them what was the one key to their success. The foundation awarded Bruce a grant and for two years Bruce spent hundreds of hours talking to these successful people. He drilled down on the one key to success. When the two years was over, he returned to the Foundation to report his results. And, what did Bruce tell the Foundation was the one key to success, risk.

Now, I'm a safety professional by trade and training and I'm not advocating a 'risk' that can get someone hurt; nor was Bruce. Bruce understood that people could misconstrue his study results and take unwise or careless risks all in the name of working toward success so Bruce put risk into categories. One category was emotional risk. I have written a definition for emotional risk, *when you do something you are a little nervous to do, that is positive and powerful, for yourself or someone else.*

If we are going to 'fix the vending machines' of our employees, we first have to know what and where these machines are. We need to be in the field coaching and giving and receiving feedback. Or, as Bruce and I like to say, taking emotional risks.

In the end, our people will grade us not on what we get done, but on how much we C.A.R.E. And, if we can care, and fix the vending machines in life, we will get results—they will be loud and clear!

16 Is Your Culture Killing You?

How to Move from a Culture of Honor to a Culture of Safety

Did you know that if you are a male less than 40 years old living in a rural part of a southern state you are nearly 20% more likely to be killed by an accident; compared to your peers from Northern states? And, this accident would stem from risky behavior. "The leading causes of death in people ages 1 through 44 are unintentional injuries, so we're looking at a substantial public health problem," said Dr. Paul Ragan, associate professor of psychiatry at Vanderbilt University Medical Center in Nashville, Tenn. Using data from the Centers for Disease Control and Prevention it is estimated that more than 7,000 accidental deaths each year are linked to this 'risky behavior.' This 'substantial public health problem' is now being called the Culture of Honor.

"Cultures of Honor," writes Malcolm Gladwell in his book, *Outliers, The Story of Success*, "tend to take root in highlands and other marginally fertile areas such as Sicily and other mountainous Basque regions of Spain. If you live on some rocky mountainside, the explanation goes, you can't farm. You probably raise goats or sheep, and the kind of culture that grows up around being a herdsman is very different from the culture that grows up around growing crops. The survival of a farmer depends on the cooperation of others in the community. But a herdsman is off by himself. Farmers also don't have to worry that their livelihood will be stolen in the night, because crops can't easily be stolen unless, of course, a thief wants to go to the trouble of harvesting an entire field on his own. But a herdsman does have to worry. He's under constant threat of ruin through the loss of his animals. So he has to be aggressive: he has to make it clear, through his words and deeds, that he is not weak. He has to be able to fight in response to even the slightest challenge to his reputation—and that's what the culture of honor means."

So why does this affect the southern states more than other states? Researchers believe it has to do with where the original inhabitants of the south came from. The southern areas of the United States were primarily settled by people coming from the most ferocious Culture of Honor—the areas of Scotland and Ireland.

In the early 1990s, researches Dov Cohen and Richard Nisbett, working at the University of Michigan, decided to test the culture of honor. After all, how can 18 to 20-year-old men living in the south still act or react as if they were herdsman? They set up a controlled scenario where each subject would be unexpectedly 'crowded' walking to an appointment. Cohen and Nisbett wanted to record each man's reaction and plot that against heritage. To more scientifically determine 'reaction', they measured facial cues, hand grip, saliva samples, testosterone and cortisol levels after the event. They wanted to know if an upper middle class male college student from Atlanta, for example, would react differently than one from New Jersey. What they found was unmistakable—the deciding factor on whether they reacted or not was where they were from. The conclusion, cultural legacies are powerful forces!

Today, our organizations have cultural legacies, deep roots and long lives. It's time to ask one basic question, do you have a 'culture of honor,' accepting risky behavior, or a culture of safety? Below are five questions to consider when you make that evaluation.

Do Small Things Matter? For southern men, who have a one in five chance to be killed in 'risky behavior,' sometimes those risks seem small. It's not wearing a helmet on the motorcycle or refusing to buckle the seat belt. In safety, small stuff matters. A number of years ago I was the lead on a workplace fatality. It was a utility case where a boom touched a 12,000 volt line and one of the workers on the ground was killed. I arrived on location about two hours after the event occurred. In looking at the scene it bothered me that one simple detail was overlooked by the crew—the wheel chock. The utility preached that when booms were in the air, chocks needed to be down. But this crew overlooked a very small detail, the wheel chock while also failing to properly control a deadly hazard, the high voltage line. Are your people taking 'small' risks? What is compliance like around wheel chocks, ear plugs, safety glasses, gloves, hard hats, seat belts? Failing to strongly adhere to these small things could mean you have a culture comfortable with risks . . . and not safety.

Is Planning Important? When I was an electrical lineman apprentice, I had one old line foreman who would take extra time to plan. He would gather the entire crew, discuss hazards, explain who was going to do what, layout the order of work, etc. He was notorious for ending each job planning session with, "And remember, nobody gets hurt today." Contrast this with what some other crew leaders did; park the trucks and went to work. After all, everyone knows what to do, there is no need to talk about it! One is a classic culture of safety and the other a culture of honor . . . which one is your organization?

Is Cowboy-ing Allowed? Have you ever heard these before? "He knows what he's doing, he's a journeyman," or, "Don't worry, he does it this way all of the time." What about this, "Don't worry about this, you should have seen what he did yesterday." If you hear comments like this, then you are in the middle of a culture of honor not safety. Research is showing that the best way to reach operational excellence is through peer to peer accountability. The opposite is going on your own, cowboy style.

Do Supervisors Own or Rent? I love the old question, "Have you ever washed a rental car?" The answer, unless you are a sick individual, is no. After all, you don't own it, why would you invest time and resources in washing it. I was recently talking to a client who was complaining that supervisors were not properly enforcing a recent training session—they were obviously just renting, not owning. A culture of safety is an owning culture . . . no renters allowed.

Is Your Workgroup a Community? "The family that eats together stays together," so the old saying reads, but I always say, "The work group that eats together is safe together." There is more to these sayings than just food. In today's busy work life, are we taking the time to build community. In farming communities community is key. Neighbors not only help each other, they depend on each other. They are not just connected by geography, living near someone, they are connected by a sense of obligation—to support and be supported. The opposite of this is of course, a herdsman's attitude. When workers feel connected, (community), they are more likely to speak up when something is unsafe, to share a near miss report and volunteer for that safety team. Is your work group a community? When is the last time you shared a meal?

The culture of honor is anything but 'honor' when it comes to at-risk behavior. Taking a short cut doesn't honor the individual who is taking the short cut nor his friends and family. Yet, like many cultures, they have just evolved. The opportunity that we have today is to intentionally start over—we can point our culture in a new direction. Leave the herd today and begin that journey to a culture of safety.

17 Hazard Intelligence
4 Intelligences that can Change Your Safety Culture

Have you heard of Chris Langan? Today he lives on a horse ranch in Northern Missouri, but the path to this ranch was not easy. Chris was born in San Francisco in 1952. As an infant, his mother moved him to Montana. Chris never knew his birthfather, and his mother remarried about the time Chris started grade school. The family lived in extreme poverty and Chris in abuse, from the step-father. The abuse continued until Chris was 14. At that time, Chris took up weight lifting and threw the step-father out of the home. Chris' high school years were spent in mostly 'independent study.' He did try his hand at college, first Reed College, then Montana State University, but didn't finish. As an adult Chris has worked mostly labor intensive jobs such as construction worker, cowboy, farmhand, and firefighter. He also worked as a bouncer in Long Island, NY. Chris' story, though somewhat heart breaking, may not be all that different from thousands of other men, yet with Chris there is one difference. Chris Langan is the smartest man in America.

Compare Chris to another 'smart man,' Albert Einstein. While Einstein may be the most noted brainiack in the United States, his life may have not been any easier than Chris Langan's. Einstein was born in Germany in 1879. As a child, his family moved several times, both within Germany and to Italy and Switzerland. Einstein moved through school, graduating from college as a math and physics teacher, but was unable to find a teaching job, so accepted a position in the Swiss Patent Office. He continued to work toward a doctoral degree then moved back to Germany to teach at the University of Berlin. He married in 1903, had three children then divorced. He remarried in 1920, but his wife died about a decade and a half later. He remained in Germany until 1933, but left his homeland due to political reasons, eventually moving to the United States.

This is where the comparison ends. Einstein, of course, has become a common name for 'smarts'. Einstein published more than 300 scientific papers and nearly 150 non-scientific papers. He earned a Nobel Prize in Physics. He was bestowed honorary doctorate degrees in science, medicine and philosophy. Einstein was asked to be the first president of Israel when the country was formed after World War II, but declined. Langan, well . . . Langan doesn't hold any degrees, has not earned a Nobel Prize, has not lectured throughout the world and has struggled to publish any of his scientific papers. If you are thinking that this isn't a fair comparison, you are right. Albert Einstein's IQ was 150 while Langan's IQ is 190—so Langan is 20% smarter than Einstein! But chances are you have never heard of Chris Langan . . . why?

Malcolm Gladwell in his outstanding book, *Outliers, The Story of Success* suggests that to be successful on the level of an Albert Einstein you need two things. The first is you need to be 'smart enough.' After this initial cut of a high IQ, it takes something more, something termed practical intelligence. Gladwell quotes psychologist Robert Sternberg, "Practical intelligence includes things like knowing what to say to whom, knowing when to say it and knowing how to say it for maximum effect." He continues, "It's knowledge that helps you read situations correctly and get what you want." Some researchers have termed this 'social intelligence'. And the best thing to understand about this concept of social intelligence is that it is teachable!

So, what does this have to do with safety? Everything! Our organizations spend enormous resources working on our safety IQ, and they should. Safety IQ includes a keen understanding and knowledge of the safety rules, safe work practices, proper standards and operating procedures. And, organizations spend considerable time studying and thinking about culture, and the effects of culture on safety results. But, we generally don't dedicate time or resources to teaching and exploring Hazard intelligence—the ability to read a job site, hazard or changing situation correctly and perform our work injury and incident free.

In fact, four key hazard intelligences, which are teachable skills, might just add up to safety success and cultural change; let's take a look:

Safety Awareness—In 1995, I began an apprenticeship in distribution line work. The training program is intense and comprehensive. It was a good combination of hands-on learning with text book modules. We had to test proficiency on a number of key tasks like chain saw use, grounding of overhead lines and termination of underground cables. What we didn't learn in the hundreds of modules was safety awareness. After all, is it okay to understand how to properly operate a chain saw, if I can't identify other site hazards that could cause a significant injury? Safety Awareness is one of the cornerstones of hazard intelligence. How well do you teach and evaluate this key skill?

Job Planning—A few years ago, as an area safety professional, I was called to a job site after an electrical contact. The worker who touched the 12,740 volt line was very lucky, he was at the hospital but would make a full recovery. In speaking to his other two crew members, I learned what had happened. The crew had energized a section of line, then took a break for lunch. During lunch, the lineworker who made contact took a phone call from his cell phone. Immediately after lunch, he went up in the bucket, completely forgetting the line was energized just thirty minutes earlier. The phone call was about his daughter, who was having a hard time. So, the injured worker's mind was not on his work. And, the crew completely failed to plan. For the most part, we teach our crews to review a job before it starts, and that's it. Jobs change and work progresses, job planning is a hazard intelligence that needs to be taught, job planning at the beginning of the job and throughout the day.

Peer-to-Peer Feedback—This year I coached my son's 9 and 10-year-old baseball team. It was a good group of players and parents, but one of the biggest challenges we had was getting the players to talk to each other. Everyone who has played baseball, softball, or any sport really, understands the importance of talking to each other; communicating. In baseball we call that 'chatter.' On our baseball team, you could hear crickets chirping . . . and many job sites and work floors are the exact same way.

In May 2010 *BusinessWeek* published an article called "The Peer Principle." This article reported research comparing organizations with good safety records to those with excellent safety records. One key finding read, "Peer accountability turned out to be the predictor of performance at every level and on every dimension of achievement. The differences between good companies and the best weren't that apparent when it came to bosses holding direct reports accountable. The differences become stark, however, when you examine how likely it is that a peer will deal with a concern." Peer to peer feedback, or 'chatter' as I like to say, is a significant key to safety success. It is also a hazard intelligence that must be taught.

Remaining Uncomfortable—Do you remember when you started driving? Or better yet, do you have a child who you are teaching to drive? We start out driving under the speed limit. Both hands are firmly pressed on the wheel. The radio is turned off, and the cell phone is safely tucked in the backpack in the backseat. We are uncomfortable and driving demands our maximum attention! Fast forward two years, the same driver has a big gulp soda in one hand and the cell phone in the other. They are driving faster than the speed limit with the radio blasting, their legs are controlling the steering wheel—they have no fear. One of the best practices to teach your organization is to stay just a little uncomfortable—staying well within safety rules and safe work practices clearly understanding the consequences of not doing so. There are a number of ways to do this, through active participation in incident analysis, near miss reports, job observations, employee sharing, etc. The key is that organizations have the hazard intelligence to remain comfortably uncomfortable . . . and safe.

In the end, IQ matters. Teach and train on safety rules and safe work practices—nothing can replace a high safety IQ. You, and your organization must make this first cut. But once the first cut is made, we need to consider what makes safety even more effective; what can help take your organization to the next level of safety performance and possibly transform your safety culture. And, just as social intelligence is a clear indicator of life success, a keen sense of hazard intelligence may just be the ticket to safety success as well. Learn, and teach well!

18 What Community can do for Safety Results

21 Things Leaders can do to Build Community and Get Results

"Give me the steak, fries and extra thick milk shake please!" Doctors and researchers over the last few decades have documented the effects on the heart of diets high in animal and saturated fat. If we were honest however, wouldn't we want to go back to those simple and tasty steak and French fry days—without the worry of any potential harmful side effects. I'm not sure we can, but there is a lesson to be learned from a small community in Roseto, Pennsylvania; people that enjoyed a steak and fries and who really weren't effected by saturated fat!

Dr. Stewart Wolf, who was then the head of Medicine at the University of Oklahoma medical center, would spend his summers in out-state Pennsylvania. One summer in the late 1950s, Wolf was asked to give a lecture to a group of fellow doctors in Pennsylvania. After the speech, a small group of doctors, including Wolf, went out for a beer. It was there that one of these doctors described a small community, a somewhat isolated town of nearly 1,700 people, named Roseto. The doctor noted in passing that he couldn't recall ever seeing anyone from Roseto under the age of 65 suffer from heart disease.

Wolf was intrigued. In the 1950s, before cholesterol-lowering drugs and aggressive measures to prevent heart disease, heart attacks were near epidemic in the United States. In fact, heart attacks were the leading cause of death in men under the age of sixty-five. This statement about 'couldn't recall ever seeing anyone from Roseto under the age of 65 suffer from heart disease' seemed hard to believe. Wolf decided to investigate.

In 1882 a group of eleven men and one boy left Roseto Valfortore, Italy and set sail for New York. Life in Italy for these men was very hard. They labored long hours in rock quarries and had received word that life might be better in America. Upon landing, they found shelter on their first night on the floor of a bar in little Italy. From there they ventured west, eventually finding jobs in a slate quarry about ninety miles west of New York, near Bangor, Pennsylvania. The next year, fifteen people left Roseto Valfortore, Italy to join the original group, then more and more. The Rosetans in America started buying land in the hillside a few miles from Bangor. They constructed homes and a church, then a park and a school. Finally, they settled on a name for their community, Roseto. The town grew. A bakery and cemetery were added along with the beginnings of traditions. A priest settled their and organized social and volunteer clubs. As the town grew, it was all Italian. In the early 1900s Italian was the language spoken in the community, so very few if any non-Italian settled there. The town flourished like this for decades, a happy community isolated from outside influences.

When Dr. Wolf arrived some six decades later he began testing and studying the residents of Roseto. And, the community readily volunteered time to his research. What he found was astonishing. In Roseto, there was virtually no one under the age of 55 who had died from a heart attack. For men over 65, the death rate for heart failure was half of the national average. In fact, the death rate from any cause was nearly 35% below rates from across the country. Wolf had to know why, so he re-doubled his efforts and hired medical students to conduct more research, interviewing every resident twenty one and over. John Bruhn, a Sociologist on the project recalls, "There was no suicide, no alcoholism, no drug addiction, and very little crime. They didn't have anyone on welfare. Then we looked at peptic ulcers. They didn't have any of those either. These people were dying of old age. That's it."

Wolf thought that Rosetan's diet must hold the secret. But he found that their mostly traditional Italian diet had 41 percent of calories from fat. In fact Rosetans didn't exercise, many smoked and many were overweight. Wolf thought maybe it was a regional phenomenon, but when he studied the surrounding communities, he found their experience of heart disease was the same as the national averages. Overtime, Wolf began to realize that the secret of Roseto was within Roseto itself. More specifically, it was the sense of community shared among Rosetans. They had a positive and proactive social culture. They valued family, friendship and helping others. People sat on front porches and visited with neighbors. People had friends. Families ate together, with three generations around the table. Bruhn later recalled, "It was magical."

Heart health research at the time was intensely concentrating on effects of genetic makeup and few if any were looking at the environmental affects that can contribute to heart attacks. What Wolf and his team discovered was groundbreaking at the time, it was the link to stress, health and community and wellness. And nearly six decades removed from Dr. Stewart Wolf setting foot in Roseto, Pennsylvania, we can still learn some valuable lessons from his research. Lessons that can give us insights on how making our workplaces more like a community, we can make happier, healthier, safer and more productive employees.

As Dr. Wolf proved through the beautiful town of Roseto, community can have profound effects on its people. Here are 21 ways that we can infuse community into the workplace:

Play Games—Raffles, give-a-ways and safety bingo are just a few examples of the games you can organize. The point is to bring people together to cheer each other on.

Display Pictures—I had a client a few years back that had a 'reasons we work safe' bulletin board. Each employee would bring in pictures of family, friends, their dog . . . with the pictures came a greater awareness of community.

Mentor—Set up a program where older more experienced employees coach and support younger employees, helping them learn the trade and safety rules.

Reverse Mentor—Set up a program where the young tech savvy employees teach the 'old hats' how to facetime with grandkids and share pictures in Snapfish . . . now that is a community building relationship.

Bring the Kids to Work—We get so locked into our daily tasks that we forget about the most important people in our lives, bringing them to work one or two days a year in an organized and safe program can build community at work and at home.

Share—Sharing a safety related story shows vulnerability, and that is how community grows.

Bring in a Speaker—This one isn't just for me, since I conduct keynote presentations. The purpose is to bring in a speaker that can make the group laugh, cry and grow together. By the way, my contact information is at the end of the article!

Laugh—Some experts say there are only two positive ways to bring a group together, through laughter and through music . . . find easy, clean and positive ways to make your people laugh.

Music—There is a high end department store that to this day pays pianists to come to their stores and play. The reason is that it gives their stores a warm community feel . . . how can you use music to bring that to your workplace?

Quote of the Day/Week—When our employees begin to think deeper, they gain perspective that helps build community and a safer worker.

Partner Coach—Who do your, or an employee of yours, go to when there is a problem or they just need advice? A partner coach is a hardwired solution.

Pledge and Track—Have each employee commit to at least two 'I will' statements, something they are going to improve on at work or at home and have their partner coach help hold them accountable to follow-through.

Start Clubs of Common Interests—There are endless after hours social clubs like, fishing, dancing, gaming, fantasy football, etc. that employees can enjoy, why not set up structures to support those? The community gained is a positive result for you.

Surprise—What can you do to surprise your workgroup? I bet you can come up with a few really good ideas.

Find a Friend—Years ago Gallop Poling Service found that employees who can identify a close friend at work are happier and more productive employees. How can you foster friendship?

Eat—A family that eats together stays together, and it is the same in a work group. How about organizing a breakfast or lunch in the next month and see what happens?

Take Walks—What about a walk around the block, or parking lot? Getting out and getting active can get results.

Greet—This may sound corny but a few years ago a utility client set up a greeting committee. Since utility workers get into a truck and drive to the work site, the committee greeted each worker at the gate as they were driving out to their respective jobs. The committee checked seat belt use, handed out sodas and candy bars and wished everyone a great day. The energy produced from this didn't have corny results!

Celebrate Birthdays, Safety Milestones, Service Anniversaries—This is an easy and 'no brainer' way to build community.

Volunteer—There are dozens of ways you and your group can volunteer time to needy community organizations. Remember, what you give is what you get back, and giving of your time as a work group will yield a great return.

Plan Family Events—Having an annual family picnic or holiday event is a good way to get started with community building.

In the end, I didn't give you any top secret ideas. In fact, your safety committee can think of a hundred additional ideas better than these. The point is simple, building community into work is an effective way to get long-term safety results. If you don't believe me, take a drive to out-state Pennsylvania and see for yourself!

19 Safety: When GOOD is Good Enough

Being good in safety is not enough. After all, the race for top decile or 'world-class' safety performance seems every bit as intense as the chase to break Roger Maris' 1961 home run record. Today, CEO's are setting single year injury goals that reduce last year's numbers by seventy five percent or more. Target zero posters hang on locker room walls all across this country. The spotlight is clearly and intensely focused on the year-end goal. Everyone is chasing zero, and that creates a small problem. It's hard to be perfect for 365 straight days!

In 2000, Dr. Kevin Leman published a book entitled, *What a Difference a Daddy Makes*. Leman, an internationally known psychologist, speaker and author, uses this book to examine the significance of the father-daughter relationship. Early on however, he takes the pressure off of us dads by letting us know that we don't have to be perfect. Leman says that we don't have to worry about being 'super-dad.' All we need to be is good. A good dad cares about his child. A good dad is trustworthy and engaged. In the end, a good dad will raise a great child. That is the truth in safety, too. If we can be GOOD (GOOD meaning; <u>G</u>=Get in the Game, <u>O</u>=Offering 3-D Feedback, <u>O</u>=One Day at a Time and <u>D</u>=Determined to Make a Difference), we can have a great safety record.

G: *Get in the Game*—A quick quiz, who do you work for? For most, the automatic response is the name on the paycheck. Others will say, 'my boss.' After some thought, a few recited 'family.' The truth is we work for ourselves. We trade time and talent to an employer for money. We are each CEOs of our own business. We each have a corporate budget, the money we have to spend, a corporate fleet, the vehicle we own and drive, and a corporate staff, our family and friends. This fact is important because if I'm hurt at work I suffer, not my employer. Sure the employer will pay a financial piece for that injury, but the great secret is that the employer will continue to make money. The injured cannot make another eye, hand or finger. Your daughter cannot make another dad; your dad can't make another son. In April 2006, the Centers for Disease Control (CDC) released a report that listed the lifetime costs of the workplace injuries that occurred in 2000; the costs topped the $406 billion mark. Those numbers include the employer paid medical expenses ($80.2 billion), but the bigger numbers ($326 billion) came from lifetime productive losses that include, "loss wages, fringe benefits and ability to perform normal household responsibilities." (CDC 2006) Those losses aren't felt by the employer but by each of our people when injured.

Recently, the National Safety Council awarded UPS Chairman and CEO Michael L. Eskew with the coveted Green Cross for Safety Medal. Eskew says this about safety, "But for safety to be a core value, it has to be taken personally." (McMillan 2007) Instilling in each of our people exactly 'who they are working for' is the first step in personal accountability and getting in the safety game.

O: *Offering 3-D Feedback*—Recently my family and I spent a week vacationing at the Disney theme parks. The most memorable attractions may have been the many 3-D shows. These are no 'typical' shows, there are smells filling the theater, rodents brushing your legs, bugs between you and your seat and a dog sneezing water on the audience. And this is just the introduction!

Today, we employ feedback mechanisms such as safety committees, peer observations, safety meetings, near miss reporting and job briefings, and these are like television sets. They are like TVs because we wouldn't live without our television sets or these feedback tools, yet often they are little more than background noise. For feedback to be effective, it has to be 3-D. We have to make it jump off of the screen and capture the attention of our people. Some 3-D feedback ideas are giving a safety committee a specific end statement. This is a vision statement but not the recipe to make the vision happen, the committee can decide what ingredients are needed for their success. Encourage peer observers to have a safety awareness item in hand. For example, a simple 'Take-Five' candy bar can be the 'attention getter' that encourages everyone to take five minutes to analyze job hazards before beginning work. Endorse Involved Safety Meeting Activities instead of safety meetings. These shift safety meetings from sit and listen to get up and do; and triple retention rates over the traditional meetings. Check out my book entitled, *Involved Safety Meeting Activities, 101 Ways to Get Your People Involved* or a book series entitled, *Games Trainers Play* for a head-start in this area.

'Practice makes perfect' isn't true; practice only makes permanence. In our business habit is key. 3-D feedback, not traditional feedback tools, is the best way to change habits and be GOOD in safety along the way.

O: *One Day at a Time*—Francis Petro, President and CEO of Hayes International Inc. said, "The fact is, the only day an employee can get injured is today. You can't get injured tomorrow until it gets here and you can't get injured yesterday because it is gone. So, we have to be very, very clearly focused on what is happening today and that becomes part of our makeup, that becomes part of our nature, and that becomes part of our culture." (McMillan 2007) Before 1997, Phillip Popovec, Site Director for International Specialty Products (ISP), said safety was, "terrible." But, the chemical manufacturer surmised, "We came to the conclusion that we don't have to worry about how many recordable injuries we get this year. We don't have to worry about how many recordable injuries we get this quarter. The only thing we have to worry about is not getting hurt today." (Smith 2005)

And, a GOOD safety program does just that. It takes the 'world-class' and 'top-decile' safety focus and pressure off the operation environment and shifts the focus to the present . . . to today. Many are engaging in Safety First meetings just like ISP. These are short, three to five minute daily safety meetings that focus on the hazards of the day and plan for making it one shift injury free. Safety stops can be used. These are predetermined times throughout a day when work is stopped for a short time to ensure all safety measures are in place. Introduce a Safety Saves program. A safety save is the stopping of a job by a peer or management person due to a hazard on that job. Once the exposure is controlled, the event is reported and celebrated. The key is finding 'what's going to hurt me today,' and eliminating those exposures.

D: *Determined to Make a Difference*—Titus Adams was a normal six-year-old boy in every way except one; Titus suffered from night terrors, a clinically diagnosed fear of the dark. On Thanksgiving Day, 2002, Titus along with his mother and two-year-old twin sisters enjoyed the holiday with grandparents. At eight o'clock that evening, his mother dressed him and his sisters in pajamas and loaded them in the truck for the hour-long drive home to Galeton, Colorado. Just minutes from their home the cell phone rang. The phone was just out of arm's reach so Titus' mom unbuckled to reach it. As she did, she inadvertently veered off the road. The truck flew into the ditch and overturned. Titus quickly surveyed the scene. He and his sisters, buckled in, were okay. Titus could hear a faint sound of a woman's voice. The voice said, "Help." Telling the twins to 'sit tight,' Titus quickly unbuckled his seat. For a long moment he peered out the broken passenger window. It was dark out there . . . so dark. Finally he took a deep breath and crawled out the window. He stood, planting his bare feet solidly in several inches of Colorado snow. The wind cut through his pajamas. It was seven degrees.

So, what happened to Titus and his family? And, what's happening to safety within your organization. There is an old saying that reads, "The road of a thousand miles begins with just one step. Take that step today." The new era in safety isn't about top-decil but a step in a new direction with intense focus on simply being GOOD. The only real question left, how determined are we to make that difference? Are we willing to do the safety equivalent of firmly planting our feet in the cold Colorado snow and walk over a quarter of a mile in the dark to summon help, as Titus Adams did. His actions saved his family. And, a quest to be GOOD can save people too . . . and create a great safety record along the way!

20 'REF' Your IFLs

Recognize, Engage and Foster Your Informal Safety Leaders

Over the last year one would be hard pressed to pick up a safety magazine and not see an article on leadership. These articles have spanned from the top (CEO) to the bottom (First Line Supervisor) of the organizational flow chart. Yet, as informative as these articles have been, we are still failing to engage a fundamental group of strong leaders within our organizations.

Leadership has nothing to do with rank or title. Leadership is when someone follows another, because they want to. A recent survey of a company with 2,500 employees asked employees to identify leaders within their organization. Only 15% of the identified leaders had a title of manager. Right now, our organizations have people leading others and these leaders carry no official status or title. These are often 'rank and file' line employees but do not let that fool you, leadership has no title. These are individuals that have great influence over the actions and attitudes of their co-workers because of who they are, leaders.

If we can identify, recognize, engage and foster this 'already existing' core group of leaders on a clear vision of safety success then we will be one large step closer to our safety goals.

Leaders, Not Managers, are Leading Your Organization . . .

Former General Electric CEO Jack Welch made this observation in his book entitled, *Straight From the Gut*. As CEO he was technically each person's boss. But most of GE's employees would never see him. "GE to them was their manager," Welch stated. He needed that manager to be walking in step with his goals and visions. But, as noted in the survey above, only about 1 in 6 managers are the recognized leaders within an organization. Subsequently, how effectively do managers establish the goals and visions of their CEO when they are managing and not leading?

When safety goals and visions are at stake, we have an option; we can engage our informal safety leaders (ISLs). Once informed of the vision and engaged on task, these leaders can achieve real change that is much harder to do from a management seat. Informal safety leaders can help improve the overall number of safe behaviors (versus at-risk behaviors), provide a means to communicate safety values and initiatives to peers, (sharing their ideas about safety with senior management), spring board their safety ideas into action when appropriate and use their already established role as leader to change the safety culture.

You'll Know it When You See it . . .

The first step in the ISL process is to identify your leaders. In reality we don't really need to define a leader, in leadership 'we know it when we see it.' Yet, just to get one thinking about recognizable leadership traits, consider a leader to have one or more of the following; an individual who takes ownership, an individual who takes great pride in his/her work, the most skilled in his/her respective trade, one who is a subject matter expert, one others listen to, one that has networks throughout the organization, one who has passion for his/her job, safety, life etc., an individual who has a positive attitude, an individual who is a leader in the community and/or an individual that has already assumed some leadership roles such as safety committee chairman or union steward. Using these criteria as a general guide, begin identifying your current ISLs.

To Change a Safety Culture, It's the People First . . .

Good To Great is Jim Collins' book outlining eleven companies that were able to sustain remarkable growth over a forty-year period. In that book, Collins noted that these companies that were able to achieve and sustain remarkable success because they realized that real change came from 'the people first with strategy, training and knowledge second.' We need to tag our ISLs, earn their buy-in of our safety vision and then tear down the walls and barriers so they can lead in their respective circle of influence. This will foster the safety change and culture that is needed for success.

After the ISLs are Identified, It's Time to 'REF' Them (Recognize, Engage, Foster)

Recognize—Many ISLs will not realize that they are leaders nor will they understand the influence they have on others by their simple actions (safe or at-risk) and words (support of the safety process). Recognizing these individuals will make them aware that they are truly leaders and have great capacity to champion (or damage) safety within their circle of influence.

Recognition can be formal, such as during a presentation, large safety meeting or company function. Or it can be personal, one on one, which is the approach I prefer. For example, after working to identify the ISLs within my organization, I recognized each with a personal hand written note and a very nice gift. The feedback was outstanding.

Engagement—After the ISLs are identified and recognized, it's time to engage them. Bring all of your ISL together in one meeting led by your company's CEO or senior Vice President. The purpose of this meeting is to clearly communicate the company's safety mission, vision, goals, strategy and expectations straight to those that are leading in safety. Learn the "Safety Pulse" (what is really going on in the field in terms of safety) from the ISLs. This gives the CEO valuable real time insight into the true safety culture of his/her organization. Communicate to the CEO the roadblocks that prevent the safety goals from being achieved. Assign specific safety projects or activities to the ISLs if warranted. Train the ISLs to take a more active role in safety leadership. Finally, encourage the ISLs to continue leading co-workers in a positive manner.

Foster—Picture a wooden wagon wheel. The wheel's center or hub represents an organization's leader or leaders. The outer rim, connected to the hub by spokes, represents an organization's employees. The failure of many wonderful companies is that in the hub there is only one leader, only one person driving an initiative (such as safety). Once that individual retires or takes a new job, your hub is gone. The wheel can't function without that hub (leader) so it collapses.

To achieve success, including safety success, it is imperative for an organization to foster leaders. If not, it is the 'captain and 1,000 foot soldiers' syndrome. Once the Captain moves on, the wheel collapses. A successful long-term safety program fosters their ISLs, moving them from the rim to the hub. This will build and establish a leadership team; a center that will ensure the safety wheel continues to turn.

Fostering the ISLs can mean meeting periodically, quarterly for example, to continue to build the relationship with the ISLs and expand on their leadership capacity. Fostering can also mean that after initially identifying a small group of ISLs, you and the ISLs recruit and foster new leaders. What starts as only a very small percent of your group (one percent for example) grows to five or ten percent . . . and the wheel keeps turning.

The Only Safe Ship in a Storm Is Leadership . . .

Picture a ship in the water. There is a water line on the side of it, some of the ship is below the water line and some is above. In taking an unusual risk, some ask, "if I try this and fail, will it put a hole above the water line or below it?" If the initiative fails yet failure means a puncture in the ship above the water line then it was a risk worth taking because the ship will sail on. If it puts a hole below the water line, then the idea is not to be tried. Failure means the ship will sink.

'REF' your ISLs. First of all it will not fail because you will be dealing with your company's leaders . . . your best people. But even if you set sail and it doesn't go the way you think it should, it's a hole above the water line; it's a true win-win. Build your leadership hub today . . . the only safe ship in a storm, is leadership.

21 The Space Between

This safety thing really makes me scratch my head. I mean you would think that our workers would perform their jobs safely, one hundred percent of the time. For one reason our workers are knowledgeable. Through trade schools, apprenticeships or other training instruction our workers are learn their job better now than any time in history. This job knowledge includes the safety rules and applicable safe work practices. Our worker's bodies are also very well trained, capable of performing precise body movements flawlessly. So, workers posses job knowledge including an expansive understanding of applicable safety rules and are physically skilled to perform their job safely, yet often the worker will cut a corner and put himself at-risk. There seems to be a space between what a worker knows is safe and what a worker actually does in the field (an at-risk act). Maybe if I can figure out what is in this space and then give it a name I can stop scratching my head while I still have some hair left.

Get Me a Flashlight, I Want to See What's Down There—That dark, musty space, the area between what a worker knows is safe and what he actually does at-risk on the job is filled with four strange things. Lurking in the crevasses of that space you will find a black lung, a by-pass heart surgeon, Kevlar and a little bit of dilly-dally. I better explain.

The Black Lung—Have you ever heard of this infamous 'black lung?' It went mainstream years ago when anti-cigarette advocates used it to convince teenagers not to smoke. The argument was simple, if you smoke your lungs will turn black and you will die. It didn't work. The reason is that people, young and old don't want to think anything bad can happen to them. The fact that we don't want to think a bad thing can happen was again reinforced by the tobacco industry when they willingly agreed to put a surgeon general warning on their product.

The tobacco industry already knew that sales would not be affected; people don't think it can happen to them.

While the tobacco industry made the term 'black lung' household, it's been around that safety industry since workers began taking risk. Workers don't think it can happen to them. Workers take risk and chose not to think about the single event that can lead to a major injury or a lifetime of bad habits that leaves the their body in poor condition on retirement day. Workers don't believe that injury or death can happen to them. They black lung it, until it's too late.

Bacon, eggs and a thick milk shake please—Dr. Dean Ornish MD, is a prominent heart surgeon, recently wrote a book entitled, *Simple Changes, Powerful Choices*. In this book he talks about his frustrations dealing with bypass heart patients. For the most part, he found that patients would come to his office feeling awful. In exchange for years of poor diet, stress and lack of exercise, these individuals had problems breathing and chest pain. After the bypass heart surgery the patient would feel great, better than he had in years. So, instead of following the prescribed diet and exercise regimen, the patient returned to the exact same environment that created the heart condition in the first place, back to the bacon, eggs, etc. What Dr. Ornish found is that this patient would be back in six months for another bypass.

Bypass surgery becomes a metaphor for addressing the symptom and not the problem. Heart patients were unwilling to alter diet and lifestyle to live a better, longer and fuller life. Safety sensitive workers are often the same. Workers, safety committees and supervisors will treat a symptom with a 'bypass' solution. Instead of addressing the need for the individual to change (and follow the prescribed safety rules) a new tool or training session will be the implemented solution, leaving everyone feeling good. However, after six months the tool will be lost, the training forgotten and the worker unchanged, continuing to work at-risk.

The Bulletproof Worker—Strap on the Kevlar, I'm bullet proof! In this mysterious space between what our workers know is safe and their at-risk behavior lurks the Kevlar or bulletproof syndrome. Workers wearing Kevlar simply think they can't get hurt. Because of this thought they conduct themselves in an at-risk manner. This mindset allows that worker to take greater and greater risk. Most of the time he is rewarded for these risks because the job is easier and quicker when done unsafe. Yet over time these risks will lead to habits. And, one morning, before work this worker will be running late because he forgot to set his alarm. He will remember his lunch box but in a rush to get out of the house and headed to work, he will forget his bulletproof vest. I send my condolences to the family.

The Two Types of Dilly-Dally—There is also dilly-dally in that space between what workers know is safe and what they actually do at-risk in the field. You may recognize the concept of dilly-dally better if termed differently, it is old fashion procrastination. Dilly-dally comes in two forms, active dilly-dally and passive dilly-dally, let's look at each.

Active dilly-dally is when a worker realizes that a work practice is at-risk and should be changed yet waits to make the change. Often the worker is consciously waiting for training or a new tool or a truck modification. He could make the change now but he continues to dilly-dally, working at risk, failing to change his behavior.

The other type of dilly-dally is passive and it is just as dangerous as active. Passive dilly-dally occurs when a worker performs a task and in the performance of this task realizes that he needs something additional or different to perform the task safely. Instead of stopping immediately and getting the proper tool or material, he continues by surveying the worksite and calculating nothing bad can happen. He continues the work, telling himself, 'I can do it with this' or 'I can make due.' He reasons that next time he will get the proper tool or material. In these cases, the worker is in the passive dilly-dally trap, putting off safety for the speed and ease of getting it done. It's a dangerous proposition because of the immediate risk and because this action might be the formation of a new at-risk habit.

Name That Space—So, just what is the space between what workers know is right and what they actually do at-risk called, it's called the safety soul. Each of us has a safety soul. After a worker is trained, skilled and knowledgeable, it's the safety soul; weak or strong that will determine the size of this space. To improve the safety soul we must turn our attention away from the catastrophic, no one ever thinks it can happen to him or her. We must turn our attention away from telling workers, you must do or else. To improve safety soul, workers must realize and become a believer in the positive benefits of safe work.

Trust Me—In closing, I want to share an exercise with you. Close your eyes. Oh yeah, if you do that you can't read this. Well, I'm going to ask you to close your eyes and point to yourself and after you have pointed to yourself you may resume reading.

So, what did you point to? Over 99% of people point to their heart. It's interesting because you don't point to your head (your intelligence, or in the safety arena your safety mind), or their hands (in the safety it's called the body or skills set), they point to the heart or the soul . . . the safety soul. We are not safe because of our heads or our hands. We are safe (or at-risk) because of our safety soul. The safety soul is who we are in safety and it determines what we will do, safe or at-risk.

22 ISMAs
Involved Safety Meeting Activities

Isaiah Thomas left Indiana University after two years and at age nineteen was the fourth pick in the 1982 NBA draft. Much to his dismay, the lowly Detroit Pistons selected him. Detroit was a team without a tradition or identity, in fact they had won just 16 of 82 games in the previous season. Isaiah was not used to losing. Just months earlier, he lead his team to the NCAA national championship. In public, Isaiah tried to put his best face forward but in private, he had serious reservations about joining such a poor performer.

But, instead of being a victim or 'just doing time' with the Pistons organization, he began studying great teams paying particular attention to what it took to build a great team. In his recent book entitled, *The Fundamentals*, he said this on building a great team; "Long-term success of sports dynasties comes when an organization is dedicated to bringing out the best in all of its people. You don't create lasting bonds with rah-rah speeches and slogans plastered on the locker room walls. You build a team by getting everyone involved from the equipment manager to the coaching staff and owner. You help every member of the organization understand how he or she contributes to the pursuit of long-term mutually beneficial goals. You establish a shared vision and a team culture with a standard of excellence and achievement. I learned that when you set the bar high and let everyone know that they are expected to push their talents to the limit at every practice and every game, your people rise to that standard of expectation."

While we're not playing basketball, building a dynasty is building a dynasty whether it is sports, business or a safety culture. We have to be dedicated to bringing out the best in our people and getting everyone involved.

The Myth of Involvement

Most leaders, managers and safety committees realize what Isaiah noted above, success is about involvement. After all, involvement equals participation, which equals commitment, which equals ownership. Ownership equals results. Yet most are confused and frustrated about how to get co-workers involved. The traditional mode of involvement goes like this. The safety leader will stand in front of his co-workers during a safety meeting and ask, beg or bribe someone to lead future meetings. When absolutely no one volunteers our leader feels defeated.

We need to turn our ideas about involvement in a safety meeting format 180 degrees. If we plan our safety meeting around an activity in which everyone participates then we have just involved the whole group. We have done so without waiting for a volunteer and without the group even realizing what has happened. This involvement in many respects is better than the traditional for two reasons; you have just involved everyone (not just a single meeting leader) and the activity is a better teacher then the traditional sit and listen safety meetings.

I Hear It and Forget, I See It and I Remember, I Do It and I Understand

Traditional safety meetings are filled with videos and reading material, all 'hear' or 'see' activities. Information is soon forgotten if received at all. Building a successful 'dynasty' comes when "an organization is dedicated to bringing out the best in all of its people" and building safety success is no different.

Bringing out the best in our people means working with your people on a level of 'understanding.' That means in safety meetings we must strive to 'do' instead of simply hear of see. "Doing" an involved activity will lead to understanding and over time, success.

The Myth of Quantity Versus Quality . . .

There was a man once, who after years of neglect, overeating, lack of exercise and basic disregard to his body, decided that he should change. He went to his doctor who performed an extensive and exhaustive physical. Two days later this man sat down with his doctor. The doctor said, "I have some good news, you're not overweight, you are exactly six inches too short." And, "you seem to be in great health for someone who is 82 years old." They both laughed knowing the man was only 45 years old. Finally, the doctor said, "all kidding aside, you weigh 300 lbs., your ideal weight is 200 lbs. I suggest exercise, eating right and reducing stress."

The man returning home set a goal to lose 100 lbs. over the course of the next two years. He was so confident that he could lose 4.16 lbs. a month; he didn't even begin dieting for the first 27 days. Two years later the man weighed 200 pounds. He was exercising and eating right and enjoying life.

When this man weighed 300 pounds, he had a lot of quantity. He lost 33% of his total body mass leaving only quality. Most safety meetings are bloated from years of neglect, lack of planning and lack of a clear goal. Each week they seem to drag on for forty minutes or even an hour, there is much quantity but not a lot of quality. In shifting your safety meetings to ISMAs, you will be focused on a clear goal and teaching through doing. In short, you will be moving from quantity to quality.

Motivational Versus Operational

Do you need operational or motivational ISMAs? You need both, and this is why. Safety can be divided into three parts, body, mind and soul, let me explain.

Body safety encompasses the physical tasks we must perform. For example, in the Utility business, an electric line worker has a physical tasks of running a chain saw, climbing a pole, operating a backhoe, installing a wood cross arm etc.

Mind safety is the knowledge we possess that allows us to do our job. It includes the safe work rules, safe work practices, standards etc.

In general, we don't take risks (shortcuts) because of a lack in body skills or knowledge. A shortcut by definition is a failure to realize ones true worth then subsequently act in a manner that places one at-risk to an injury.

In short, we need both operational and motivational ISMAs. Operational activities will re-train on safety rules and safe work practices. Motivational activities are aimed at increasing worker motivation, understanding of values, self-worth and sense of family so that they will not place themselves or co-workers at risk.

The Nuts and Bolts . . .

For operational ISMAs, break down a safety rule asking these questions, how can we do it, solve it or ask questions about it? Using an example from the utility industry, one rule reads, "Before setting a meter, the worker shall ensure, proper voltage, no back-feed or grounds in the meter base." Traditionally, the safety meeting facilitator would review this by reading the requirement and then verbally reviewing the proper tests to be performed to ensure compliance, a real sleeper. But, if we plug in a question above, "can we do it?" we realize there is a better way. Breaking the group into smaller groups of four, we can bring in a meter base for each group and have them perform the proper checks while wearing all of the proper PPE. Instead of reading to them, each person has the opportunity to review it 'hands on;' talk about understanding!

For motivational ISMAs, check out my book, *ISMA—Involved Safety Meeting Activities: 101 Ways to Get Your People Involved* or try an Internet search using key words 'games trainer's play.'

It's Hard Simply Because It's New!

Isaiah Thomas discovered making a winning team took hard work and determination but over time he did it, winning two NBA championships with the Detroit Pistons. His secrets of bringing out the best in people and getting everyone involved are the same key concepts at the heart of each ISMA. The ISMA concept isn't hard, just different. Hard work, determination and the willingness to make this small change will change your culture forever . . . let's play!

23 Is Technology Hurting Our Workers?

Will Change Management Evolve to Address the New Culture of Technology?

Is technology hurting our workers? Before you answer, take this two question quiz. The first question, what does "MOS" mean? Next, do you know the meaning of "PAW?" If you are 'hip' on the new language of text messaging and/ or have teenage children, then these are easy. On the other hand if you are like me this newly emerging language makes absolutely no sense. By the way, MOS means 'mom over shoulder' and PAW is 'parents are watching.' This is a simple example of the changes being driven by technology.

Of course, change management is nothing new to safety, especially if you work in a System Safety environment. Each new piece of equipment or process improvement is analyzed for hazards and ultimately for any negative effects on the system and the safety of our people. In fact, Joe Stephenson in his book, *System Safety 2000* wrote, "Change is a necessary ingredient of progress. It can be a positive factor that improves the effectiveness, efficiency, and /or safety of an organization or operation. However, change is almost always a causal factor in accidents, many times a very significant causal factor. Whether change is a friend or foe depends largely upon the manner in which changes are planned, managed, and controlled."

Today, technology is fueling a fundamental shift, or change. And, it's not just a change for teenagers pushing buttons on cell phones. These external changes are seen everyday, for example, card scanners at the grocery store, hi-speed internet, You-Tube, I-phones, I-pods, Blackberry's, internet at one's finger tips, and the list goes on. The point is that these changes are having a profound effect not just on college campuses, libraries and homes across the county, but these changes are transforming the attitudes, habits and culture within our work environments. And, when the work environment is changing, that means workplace safety will be affected in some form or fashion. The following are four examples where this technology will have an impact on our work environments.

Can You Hear Me Now?—We communicate much differently today than two decades ago and much differently today than even a few months ago. Can you remember when you saw the first cell phone being used in public or the first Bluetooth attached to someone's ear or the first IPod strapped to the bicep? Those technologies are common place both in public and the workplace. Fueled by technology, the face of communication is changing rapidly in at least three primary directions. First, it's much harder today to communicate without distractions, whether it's someone leaving a meeting due to a cell phone call or hearing another person typing as you converse over the phone, technology means more interference in conversations. Technology means fewer communications are face to face as employers are providing employees with email, Blackberry's and cell phones. It's easier, yet far less effective, to handle a critical conversation over the 'net' instead of face to face. Last, in the safety world, many applications require employees to 'follow the procedure;' yet technology is fueling a 'social media' where any reader can become a participant to share a comment or feedback with the click of a mouse. Today's workers are less accepting of 'do it because' procedure. They want a chance to click in for comment and answers.

Possessing the Attention Span of a Gold Fish!—If you have made it this far in the article, you have already surpassed a growing number of people suffering from 'grasshopper syndrome.' You have been able to ignore the television, email, computer, Blackberry, etc., and beat the grasshopper syndrome, which is the urge to quickly jump from one thing to another within the span of nine seconds. Our attention spans, and those of our workers, are key to success in safety. In fact, Dr. Ken Gibson in his book, *Unlock the Einstein,* says, "Attention gives us several ways to constantly and appropriately monitor our environment."

Technology, however, may be eroding attention spans and thus that key ability to monitor the work environment. A report published on February 22, 2002 stated, "Our attention span gets affected by the way we do things," says Ted Selker, an expert in the online equivalent of body language at the Massachusetts Institute of Technology. "If we spend our time flitting from one thing to another on the web, we can get into a habit of not concentrating," On February 22, 2002 The British Broadcasting Company added, "The addictive nature of web browsing can leave you with an attention span of nine seconds—the same as a goldfish." Attention spans, and lack thereof, can affect traditional training, instruction, work periods, breaks, etc. I'd like to expound on this but it's been over nine-seconds . . .

A New Name for the Car: 'Mobile Phone Booth'—Today's notion of multitasking is fueled by technology. For example, today we seem to use our car for everything but driving. Much has changed about automobile travel since Henry Ford said we could have any color vehicle, "as long as it is black," or since the first cup holders were installed in the 1986 Taurus. I realized just how much has changed the other day when I was driving to a meeting. Being the safety professional, I had the cruise control set at an appropriate speed, enjoying the drive. I noticed a car approaching quickly from the rear then moving past me. As his car inched by mine, I noticed the driver was typing on a blackberry and sipping coffee. I began to honk. I wanted to send him a signal to pay attention to the road. Before I depressed the horn I realized I knew this guy . . . he was a co-worker! In reality, what I saw is typical. Today, cars are for breakfast, dressing, shaving, texting, cell phone talking, the application of makeup, reading, dinner and more texting. This shift of multi-tasking while performing a safety sensitive task (driving) is not limited to the car. Today, as I see workers sporting Bluetooth ear buds while in the course of task completion or checking email from a Blackberry at lunch, I wonder just how far this multi-tasking habit has inched into our workplace.

Transparency—Sitting in many business management meetings over the last couple of months, it seems that transparency is the new buzz word for high paid consultants; bubbling ahead of past favorites such as execution and accountability. While buzz words will come and go as consultants search for billable hours, they just may be onto something with this new word. According to Paul Leinberger, founder of TDI Consulting (Trend-Driven Innovation), transparency is the result of today's advances in technology. In a recent presentation to a utility company, Leinberger said that today there are no secrets. With over 2 billion people world wide with cell phones, many with cameras, nothing goes unnoticed. When captured by camera, the image can be sent world-wide in a matter of seconds. To that end, 'spin' is coming to an end, with truth being the only 'truth' left. Within our organizations, how transparent are we? Are we 'real time' with our information? Do we keep our employees in the loop? Failing to adjust our style, pace and communications whether we think it's important or not, can foster anxiety and stress within an organization. How transparent are we?

So, is your organization G2G (Good to Go)? What type of safety hazards might be lurking due to the shift in technology? Remember, "Whether change is a friend or foe depends largely upon the manner in which changes are planned, managed, and controlled." How we anticipate and respond to the technology revolution both inside and outside our workplace will have an effect on our success in safety.

24 From the Inside Out

My heart immediately sinks. Even the most severe injury in our company in years doesn't produce a change in work habits.

I supervise electrical distribution line workers. We start each week with a safety meeting. Being the highly effective supervisor that I am, I have a folder of safety stuff I draw from when we don't have a video or speaker.

On this day, I reach into this stack and find some real gems. I pull a recent incident summary where a truck overturned. I snag a near miss resulting from a defective tool. On the bottom of this stack, I slip in an injury update. Three months ago, a line worker in our company was hit by a car as he and his crew were working along a road. This particular update talks about his continued recovery. I surmise this will be a great safety meeting closer, leaving a true impression.

The meeting begins and I work through the incident summary, and then the near miss, answering questions and getting some good comments as we go. With all questions answered, I go into a dramatic mode. I want to send a message and leave them with "safety" fuel for the week. The injury update talks about how the worker is being moved to a new rehabilitation unit. While he's there, one of the first activities will be throat exercises. The goal is to strengthen his muscles enough to remove the feeding tube.

It's sobering. I let the last words hang in the room. I close the meeting with "be safe."

After the meeting, it's the usual stuff for me. I hand out construction jobs, answer questions and phone calls. Later that day, I escape the office to see how the crews are getting along. I pull up on one crew and my heart immediately sinks.

There, in the middle of the road waving traffic, is one of my workers. I didn't see any traffic signs as I approached. The crew doesn't have cones or other traffic protection in place. And, to top it off, the worker in the middle of the road isn't wearing his orange flagging vest. Wasn't he in the safety meeting this morning to hear what can happen if he's hit by a car? Why hasn't this incident, a co-worker's serious injury, changed his behavior? What happened to "be safe"?

EMEs: External Motivating Events—A serious injury to a co-worker is an example of an external motivating event, or EME. An EME is an event or action, whether planned or unexpected, that happens to an individual. Think of an EME as "doing" something. The safety meeting is an event that happened to my group on that particular morning—it happened, they experienced it.

EMEs can be on a national level, such as 9/11; on a group level, such as the safety meeting; or on an individual level. Other examples of EMEs are the birth of your child or grandchild, securing a new job, purchasing a new home, or a near miss on the job. The list is endless.

EMEs have a profound impact on lives. They leave impressions so great that they produce a commitment to change. Yet because that commitment was motivated from the outside in, it generally is a temporary change, if done at all. Even the most severe injury in our company in years doesn't produce a change in work habits.

Externally motivated events are the basis of most safety programs. Today safety processes are based on creating events for workers. We call it "doing" safety. There is a whole host of things we do for safety, such as safety meetings, incident reports, safety audits, equipment inspections, training programs, and review classes. Each EME is designed to change behavior from the outside in.

The problem with motivating safe behavior with external forces, however, is twofold. First, as a rule, people temporarily change (if they do change) from external cues and quickly revert back to the old at-risk behavior.

Motivating with external events or "doing" safety is becoming the measure of safety performance in many companies. Generally, though, only some in the workgroup actually do safety. And fewer of those doing safety even do activities that would motivate a shift in behavior. The vast majority stays with the status quo, habits and work practices that are generally less than 100 percent safe.

With a few of the group in the mode of do, do, do and the company masses turning a deaf ear, we simply end up with a pile of do-do and the same old (or worse) injury log. The poor record is hard to digest, however. Along with upper management, we are left holding our noses and wondering why we continue to have such a stinky safety record.

The Internal Motivating Event—Now, don't throw this whole idea down the toilet. It's not our fault that we base safety on EMEs or "doing." We are people, and it's human nature to do things. We are programmed at a young age to ask, "What can I do to help?" Next, our companies are liable and responsible for ensuring workers conform to and practice established work rules. As employers, we have to establish programs (doing) that ensure safe work.

But—and it's a big but—lost somewhere in the need to do and do and do so much for safety is the simple and lasting power of the internal motivating event, or IME. An IME is the independent choice of an individual to shift thinking and in so doing, permanently change behavior. We can think of IME as *being* instead of *doing*. We see IMEs daily when the 30-year smoker decides to quit, or an overweight person changes eating and exercise habits to keep the weight off, or an individual commits to seat belt use.

Moving to a State of Be-ing—So, what does it mean to say "be safe"? I think it means we do safety (EME) by following all applicable rules for each task. And, it goes beyond just doing, it goes into be-ing. I want them to make an internally motivated decision to act in a manner where they're be-ing safe in each act and not just doing it safely.

The old adage says, "As I be-lieve, so I be-have," not "As I do, so I believe." Actions may change based on external conditions. A state of be-ing will not waver.

In all of our doing for safety, we need to spend some of our resources attempting to change behavior from inside out. We should remember the lasting power of an IME and the temporary (if at all) shift of an EME.

Telling a worker to "be safe" takes root only when safety is motivated from the be-ing. The secret to safety success lies not from without (outside), but from within. Be safe.

25 Are We Too Comfortable?
WAKE-UP and Snap Out of the Comfort Zone

"If I could wave a magic wand," said Ray LaHood, Secretary of the Department of Transportation, at a recent Distracted Driving Conference, "I would eliminate it, but I don't have that magic wand." LaHood was talking about the epidemic of unfocused drivers on this countries highways and interstates. More specifically, he was talking about texting while driving. La Hood continued, noting that it "seems to be getting worse every year."

Passengers on Northwest flight 188 from San Diego to Minneapolis wish they had had a magic wand during an October 2009 flight. Their pilots, for some reason, flew 150 miles past their destination, the Minneapolis airport. On the ground, air traffic controllers became very nervous when the plane's pilot and co-pilot failed to respond to repeated radio calls. In fact, two Air National Guard jets were notified and placed on stand-by. Finally, the pilot responded. After a thorough investigation, it was found that the pilot and co-pilot were on their laptops, talking about work scheduling, not paying attention to the flight plan, aircraft or the safety of their 144 passengers!

There seems to be a question just below the surface of texting while driving and laptop use in the cockpit; are we becoming too comfortable? Operating a car while eating, drinking, talking on the cell phone and texting seems to be more common every day. While these pilots were caught with the proverbial 'laptop open' this certainly isn't the first incident, just the first one making national headlines. A few months back, an engineer of a passenger train was texting; his inattention caused a major crash. It seems there are two distinct trends coming together that make these distractions easy.

First, there are no longer great discomforts while operating equipment or machinery. In years past, it was a 'major' job to operate equipment and machinery, yet today we have power steering, climate control, GPS, autopilot, etc. These comforts have somewhat removed us from the reality that driving 75 mph or flying a plane is still a safety sensitive task! Next, we have a major shift in technology. Five to seven years ago we would not have had a conference focused on eliminating texting nor would we have had pilots with laptops. Today, we have blue tooth earphones, iPods and 4G networks that expose us, and our workers, to continual distractions. The comfort and perceived safety of today's equipment and machinery coupled with this technology advancement is trouble. We need to snap out of this comfort zone . . . we need to WAKE-UP!

The WAKE UP model is a simple and practical approach to snapping out of the comfort zone and focusing on the task, and hazards, at hand.

W: *Watch for Distractions*—The first step is to survey the work area, driver's seat or cock-it. This is the simple, tried and tested workplace analysis where we canvass the scene to determine, what can hurt me here? In the case of being too comfortable, what can cause distractions?

A: *Ask Questions*—And encourage a questioning attitude. Over the last few years, gadgets have entered the workplace and the pace of new technologies will only accelerate. As we are Watching for Distractions (surveying the scene) we need to question everything. Does an item distract or pull attention away from the job? Does taking my eyes off the road, even for a second, put me at a greater risk? Can I perform the task for 30 minutes without my bluetooth in my ear? Can I talk about scheduling another time, maybe after I land the plane?

K: *Know the Results of Being Too Comfortable!*—Reggie Shaw was 19-years old when he hopped in the car for a quick errand. "I thought it was safe," Reggie recalls, "I thought it was something I could do. That I could drive down the road and send a text and be safe." Reggie, while sending a text hit another car, killing two passengers in that car. He landed in prison for a month and now speaks out on the dangers of driving and texting. Most of us having worked in safety sensitive industries for some time can recall a host of incidents, accidents, injuries and unfortunately, probably even a fatality due to lack of focus. We know what the results can be . . . let's not forget!

E: *Enforce Rules*—As information about the Northwest pilot's activity was hitting the news; many airlines began reporting their rules regarding laptops in cockpits. Some airlines do not allow laptops in cockpits at all. Some don't allow them while the plane is in flight. Many states are passing laws against texting and driving. Some states have rules restricting cell phone use in cars to hands free devices only. In short, rules are important steps that tell the entire workforce that there is a recognized hazard, the results of not eliminating the hazard are great and therefore, this rule is established as a minimum level of safety.

U: *Understand It's More Than the Rule*—"Our people don't care what we know," an old quote reads, "Until they know we care." In short, rules are the first step, but an equal if not more important step is one of trust and understanding. Josh McDowell, author and family expert says, "Rules without relationships causes rebellion."

P: *People Make It Successful, Not the Rule*—Just as we are talking about rules and relationships we have to remember that we don't manage safety, we manage people. How we communicate, educate and enforce, if needed, will be the key to any successful shift in practice. Maybe LaHood said it best during his opening remarks at the Distracted Driving Conference, "I want to remind everyone that we cannot rely on legal action alone because, in reality, you can't legislate behavior." Taking personal responsibility for our actions is the key to all this."

How comfortable are you and how comfortable are your people? Or maybe a better question, who in your organization is performing a safety sensitive task with an iPod, blue tooth, cell phone, text messaging system or laptop alongside? Comfort does run out!

26 Fire Your Consultants and Do It Yourself, FREE!

Performance Excellence Through the Three Building Blocks of Safety

I have a nightmare that I'll have to break my son's heart. While he is only six-years-old now, the dream goes like this. Immediately after college, looking so proud in his cap and gown, he runs to me and confesses that his dream is to work for Southwest Airlines. Immediately following his celebratory graduation dinner, he will submit his brilliant credentials to the airline for consideration. He hopes to start at the ticket counter or as a baggage handler then work his way up. As a father, my heart drops and I have to respond. "Son," I begin, "Don't you think you may be setting your sights too high? Why don't you come back to earth and do something more reasonable like apply to Harvard Business School!"

It's true-statistically one has a better chance of admission to Harvard Business School than to gain employment at Southwest. This airline has won five consecutive Triple Crown awards for best on-time arrival, best baggage handling and fewest customer complaints. And, they did it through culture. "The culture of southwest is probably it's major competitive advantage," Herb Kelleher founder and former CEO says, "the intangibles are more important than the tangibles because you can always imitate the tangibles. You can buy the airplanes and rent the ticket counter space. By the hardest thing so someone to emulate is the spirit of your people."

The problem is that most of our organizations only wish we had a culture like Southwest's. Instead we are average-to-good with customer service, mediocre with product delivery and are 'on time,' some of the time. Add to that, it seems much harder to 'fix' the problem in today's tough economic climate. There is good news, you can do it and save money along the way. First, your consultants and then do it yourself-free . . . through safety. Performance, operational excellence and a sustainable culture can bring talent to your door; just ask the 90,000 who applied to Southwest last year. And, it can all be achieved through the three building blocks of safety.

Appreciation and Recognition—Steve Chandler in his ground breaking book, *100 Ways to Motivate Others* introduced the reader to Professor Mercado. Mercado was a violin protégé, college professor as well as genius in numerous fields including math, economics and music theory. He was also a genius in appreciation and recognition. One example was when the parents of a boy named Michael asked him to teach their son violin. Michael however wasn't a typical student. For starters, he wore his hair long, and straight down over the face. One couldn't see Michael's eyes or facial expressions. In addition to the appearance, Michael didn't speak much. Week after week Professor Mercado would teach the violin to Michael who would sit in silence. It was strictly a one way communication, not unlike some of our employee-employer relations. At times Michael would not even pick up the violin. Although receiving no feedback from his student, no sign of progress or success, Mercado continued to teach, focusing on Michael's potential within . . . the skill and talent hidden under the hair and silence. Finally, one day when Michael was in the eighth grade he picked up the violin and began to play. And, in less than a month he was asked to solo for a regional symphony.

"Recognition," says Richard M. Kovacevich CEO of Wells Fargo, "is American's most underused motivational tool!" While Michael's story may be extreme, it shows the supreme power of appreciation and recognition. In fact, the insightful book entitled, *The Invisible Employee* reported some remarkable statistics on the subject. "According to a 2003 survey, 90% of workers say they want their leaders to notice their efforts and improve their recognition and rewards." In addition,

"In an ongoing Gallup survey of more than 4 million employees worldwide, there is remarkable evidence of the business impact of recognition and praise. In a supporting analysis of 10,000 business units within 30 industries, Gallup found that employees who are recognized regularly increase their individually productivity, increase engagement among their colleagues, are more likely to stay with their organization, receive higher loyalty and satisfaction scores from customers, have better safety records and fewer accidents on the job." Safety is tailor-made vehicle for appreciation and recognition. It is in the natural facilitation of safety process elements such as job safety observations, safety team interaction, and safety goal achievement where consistent appreciation and recognition can grow roots and have a far reaching positive affect on an organization.

Support—In early December of 2006, Mid-Missouri was blanketed by nearly 24-inches of snow. In an area of the country where a half-foot of the white stuff can close schools, 2-feet immobilized the region. In addition, many warehouses and factories were damaged when roofs collapsed under the weight of the snow. One of the more intriguing roof cave-ins however was a large horse barn. While most roofs failed in the first 24-hours, this one failed five days later. After an investigation by insurance officials, it was determined that the roof was built to withstand the 24-inches of snow, when that snow was distributed evenly. There were, however, several periods of melting and re-freezing and the snow shifted, sliding and gathering along the mid-point on each side, causing failure.

"Every organization," Stephen Covey says, "is uniquely designed to exactly produce the results it achieves." Just as the roof was designed to achieve certain results under certain circumstances, our organizations are much the same. If we ever wondered why your teams fail or our people under perform-remember it is about simply design-that is how our 'roof' is designed. We can climb out however through safety. How do we support the safety process, our safety teams, incident analysis, process change management, etc.? The best way is to ask and continue to ask, at all levels, what support is needed? Offering support, first through safety can allow our organizational 'roof' to stand strong over many storms.

Critical Conversations—Mary Kramer and her family had a new neighbor, Bella. Bella wouldn't have been their pick of neighbor's but early on they decided to tolerate the pit bull puppy. As the dog grew, it never really caused trouble, Mary recalls. Sure it would eat her dog's food and chase her horses but what could she do? In Boone County, Missouri there wasn't a leash law for dogs. Mary thought about talking to the owners but stopped short saying, "You never know how somebody's going to take it. I don't want to be confrontational." Last week Mary's ten-year-old son Tyler was walking to the bus stop. From behind he heard the sound of 'steps on gravel'. He turned to see Bella in mid-air lunging toward him. Tyler knew that if attacked by a dog or other large animal that one should curl up in the fetal position guarding the head and neck. That's what Tyler did, he forgot however about the last half of that instruction, to remain quiet, if possible.

Curled up, screaming, Tyler laid in the grass of the home next door as Bella ripped his backpack and jeans to shreds. Mary heard her son's screams and immediately came running outside and charged the dog. Luckily the dog ran off. Mary picked up Tyler returning to the safety of their home to call authorities.

The old wise quote from an anonymous author says, "An excuse is just a reason packaged with a lie." Sometimes on our jobs we stop short of talking to a co-worker about safety for the same reason as Mary Kramer, "You never know how somebody's going to take it. I don't want to be confrontational." Performance and operational excellence can only come through thousands of critical conversations and the best starting point for these can be in the field or on the floor, where safety sensitive work is being done. Caring about others means that we have these critical conversations at all costs. In the end, Tyler was okay through luck . . . we can't bank on that luck within our organizations.

Can appreciation and recognition, support and critical conversations beginning through the window of safety really propel ones organization to excellence? Yes, it is that simple. For example, when Paul O'Neil took over Alcoa in 1988, he knew nothing about aluminum, just that the company was in big trouble. It was national consensus that it would soon be impossible for an aluminum manufac-

113

turer like Alcoa to make a profit in the United States. Outsourcing to other countries was the future. In fact, the previous CEO had begun to purchase other business, hedging future losses and preparing to close the aluminum operation. O'Neill disagreed, however, but understood the path to performance and operational excellence would be difficult, at best. Margins would be thin, super thin. The key to success was through the safety of Alcoa's 145,000 employees.

"On his first day, he told Alcoa's executives that they weren't going to talk people into buying more aluminum and that they weren't going to be able to raise prices, so the only way to improve the company's fortunes was to lower its costs. And the only way to do that was with the cooperation of Alcoa's workers. And the only way to get that was to show them that you actually cared about them. And the only way to do that was actually to care about them. And the way to do that was to establish, as the first priority of Alcoa, the elimination of all job-related injuries. Any executive who didn't make worker safety his personal fetish—a higher priority than profits—would be fired. (*New York Times Magazine,* January 13, 2002 by Michael Lewis).

Did this focus on safety work? Yes, in the next 12 years, he doubled Alcoa's global market share and more than doubled its number of workers. After several years of depressed earnings in the early 90s, O'Neill took Alcoa from a profit of $4.8 million in 1993 to a profit of $1.5 billion in 2000—in other words, performance and operational excellence through safety.

In the end, we will probably not fire the consultants nor will we transform our organizations. But, if some day appreciation and recognition, support and critical conversations become the center piece of safety-the window to performance and operational excellence will be opened. And when that happens, I might just be telling my son not to dream about working at your organization either, instead to apply to Harvard Business School.

27 Finding the Smoking Gun

The Five Things that No One Will Talk About in Near Miss Reporting, and Questions to Foster an Effective Reporting Environment

There is no doubt that near miss reporting is important . . . very important. As safety professionals and leaders, we know that a free lesson learned today is an injury avoided tomorrow. And, there are dollars associated with near misses-serious dollars. If we ignore the tremendous 'human' and emotional impact of serious injuries and fatalities and only compare their financial costs with the costs of near miss injuries, it might be an eye-opening exercise. According to some estimates, near miss events may cost twice as much as serious incidents or fatalities. According to a Houston Business Bureau, CII and Exxon Chemical report, a near miss event is estimated to cost about $1,300 and they estimate about 1,000 near miss events for every fatality. Using 2004 Bureau of Labor Statistics data, 5,703 workplace fatalities were reported across the United States. At an estimated million dollars per fatality, near misses cost the private sector more than a trillion dollars more, on the monetary side of the equation, than fatality. And, if estimated costs are shifted, near misses can move this ledger amount to twice that of fatalities.

As safety professionals, we already agree about the importance of near miss reporting. Add to this foundational understanding a cost basis rational, then a question flows naturally-why don't we encourage reporting and thoroughly analyze near miss events? After all, how many near misses have you worked on this week, or this month? Chances are, you can count them on one had . . . one finger probably. In truth, there are five key reasons why these events are not reported and analyzed. Five truths that no one wants to talk about, until now.

Unspoken Truth I: *They Aren't Free!*—Any article or training class dealing with near miss reporting will frame a near miss event as a 'free lesson' or a 'golden nugget' that must be analyzed for learning. The truth, however, is that these events are not free! If you are still skeptical, then tell me how many near miss reports have come across your desk in the last week . . . I rest my case. In most organizations there is a cost to near miss reporting. That cost can come in many forms such as loss of credibility by the worker reporting the near miss. The price might be in an intimidating reporting system or unspoken signals from line-management. The cost might be in the perceived time and hassle to make the report. Our critical role is to identify these costs and work to reduce them.

Questioning Attitude—What are the hidden costs to near miss reporting within your organization? How can we put ourselves in the shoes of our workers to truly understand these costs?

Unspoken Truth II: *Free is Not Enough!*—In the last paragraph, we just established that we must find the hidden cost of near miss reporting then blow them up, eliminate them . . . after that we must confront the next truth that no one is talking about . . . free isn't enough! There has to be an incentive or motivator in order for our people to engage in a near miss reporting process and work an incident through the process. Put yourself in the place of your workers. It's a typical afternoon, and things seem to be going fine. All of the sudden, bam! The worker is almost injured due to a part defect. She quickly replaces the defective part and begins the task, as she does this she realizes that she was lucky. Had she not noticed the defect, there is a great chance she would have been injured. At that moment, the worker thinks . . . I wonder if I should report it? The thought that comes next is key . . . will she be motivated to do so or not? In most cases, free isn't enough, there must be a positive motivator established so our workers will answer the 'should I report it' question with a strong positive.

Questioning Attitude—Is there motivation to report near miss events? At the moment a worker realizes a near miss occurred, what can be the motivating factor to allow him/her to report the event? What is the 'right' type of motivator for your organization? Should I overcompensate with a motivator in the short term to establish a long-term habit?

Unspoken Truth III: *It's Too Much Work*—Okay, let's say that your organization is above average and near miss reporting is free, in fact there is motivation . . . there is another hurdle that no one talks about . . . line-management doesn't want the added work! If we refer back to our college days and reference the great H. W. Heinrich's Accident triangle from his book, *Industrial Accident Prevention: A Scientific Approach* we recall that for every major injury there are 29 minor injuries and 300 near miss events. In looking at these ratios in one's organization, it is probably safe to say that if reported, a line-manager could have at least one, if not more, near miss events a week. That being said, how busy are your line-mangers right now? Chances are they are swamped. What are they going to do with another one or two near miss report events per week? Chances are they won't do anything with them . . . and that's the fear! By the way, in all honesty, what would your organization think of a supervisor who managed over 100 near miss events in a year? While safety professionals would think it's great, do you think managers would perceive an issue with the supervisor's performance?

Questioning Attitude—What kind of near miss reporting system can speed up the process, be effective in eliminating future exposures and assist line-managers at the same time? How can line-managers be rewarded and motivated to follow near miss events through the process? How can upper management recognize a high number of near miss reports as 'effective supervision' instead of 'poor performance' by a line manager? Does an organization need a separate position just to manage near miss events?

Unspoken Truth IV: *No One Knows What a Near Miss is!*—One of the most effective football coaches of all time, Vince Lombardi, used to say, "It's hard to be effective when you are confused!" That same spirit translates into near miss reporting. The fact is that our people have a hard time defining a near miss and have a hard time translating a field event to a near miss report. And, when there is uncertainty, people will withdraw into silence and status quo.

Questioning Attitude—How can we design a near miss definition that is both effective and easily understood? How can we evaluate our employee's current understanding of near misses?

Unspoken Truth V: *The Process Stinks!*—How do you feel about your annual trip to your state's license bureau to renew your driver's license? Chances are that you would rather have a root canal than stand in that line for forty-minutes only to talk to an angry license bureau employee and find out you didn't bring the right paperwork! In the eyes of most employees, near miss reporting is the same type of event. Think about it, something happens and the result is that no one gets hurt, yet the employee or crew is dragged into a room with management for an 'investigation.' After a couple hours of seemingly pointless questions that meeting ends, few, if any, changes are made . . . what' doesn't stink about that?

Questioning Attitude—How can technology be used to help the process, how about a near miss blog? How can the process be turned from 'stink' to roses? What if employees ran the process with a fostering eye and consistent support from management?

Having worked over a decade and a half in the utility industry it is safe to say that the most dangerous work for utilities workers is during storm restoration. After a storm, weather conditions are generally poor and electrical hazards extreme. To help mitigate these hazards and maintain a high level of safety awareness, crews will gather every morning to review hazards and then meet again in the evening to review the day. It's in these evening sessions that line workers will openly share near miss events. It's a free exchange motivated by a general intent to keep a 'brother or sister' from being injured by the same or similar exposure. The process is supported by line-management and safety staff alike, it is free, workers are motivated by a sense of genuine caring and a desire to help, it's not full of process or forms and it is done in a timely manner. Yet, after a storm, most of this free sharing is lost as crews revert back to 'normal' hazards and office politics.

Near miss reporting is one key to safety success, yet lost in most organizations. In the end, it's up to us as safety leaders and professionals to speak the unspoken truths and make a process where experiences are shared and supported for one common goal, to eliminate that human side of loss—before it happens!

28 Safety Stop
Using H.U.R.T. to Eliminate Incidents and Injuries!

It was a stormy evening when the phone rang. The wind and ran had been pounding the windows of the rental for nearly 30 minutes and I knew exactly who was calling. I finished the phone call and put on my boots, I was going to work.

I learned that we had several hundred people out of lights. I was given a handful of orders and began to work them, going to homes to check meters and services; clearing limbs and re-fusing cutouts. I was feeling pretty good about things. I was a new journeyman, so working trouble by myself, something that I would not have done as an apprentice, was feeling good; empowering. In the middle of one order the dispatcher called. The 34.5 kV circuit that fed three small towns had locked open, I needed to go to the substation for switching instructions.

By the time I arrived at the substation, evening had turned to night. The temperature had dropped another 15-degrees and the rain continued to pound. In talking to the dispatcher, we needed to open the breaker and associated switches so we could do some line repair. After the repair, we could close the switches and breaker restoring power to over 1,000 customers. In all, there were three instructions and I accomplished the first two without incident. The final order, to check open a certain switch became a problem because I could not find the switch number. I looked everywhere. I began to hurry and became a little frantic. I knew the dispatcher was waiting for me, so that a crew on site could begin work once I finished these orders. I was irritated and frustrated that I could not find the switch location. Finally, I found it, there was no switch handle. I shined my flashlight through the heavy rain and substation steel to see the switch, I could barely see it. I was in a feverish pace at this point, looking everywhere for a switch handle so I could operate the switch.

Finally, after what seemed like forever, I found a handle, inserted it into the switch and operated the switch. It was about half way through that procedure when something just felt wrong. As I finished the operation, I immediately knew that I had closed the switch, which was exactly what I was not supposed to do. Because I was hurrying and annoyed, I had just closed the bypass switch, reenergizing the line. Without thinking, I immediately threw the switch open. If you are a lineman, you know exactly what happened; a huge fire ensued because I dropped the load of three small towns with a solid blade switch!

In the end, relays at a bulk substation put the fire out. No one was hurt and I was extremely lucky that no equipment was damaged. What I didn't know then but understand now is that I broke two of the four H.U.R.T. laws; and if we can recognize when we are in H.U.R.T. mode and perform a safety stop, we can go a long way to preventing incidences like I just described and many injuries, too. H.U.R.T. stands for;

Hurrying—Legendary collage basketball coach John Wooden used to tell his players, "be quick, but don't hurry." What Wooden understood, is that when his players would hurry, they would lose sight of a bigger picture and make costly mistakes. They would become single focused and loose sight of other players, teammates and the play the team was trying to run. If we find ourselves hurrying on a job, stop! It's okay to be quick, thoughtful, but if we are rushing around, perform a safety stop.

Upset—A few years ago, working as a safety professional for a mid-western utility, I received a call that a lineman had contacted 12kV. I immediately drove to the location and interviewed all involved. The crew had energized a section of line then took lunch. Immediately after lunch, one of the crewmembers went back up in the air to make jumpers, completely forgetting the line was energized. He was lucky, that he didn't get seriously hurt. What we found, however, was that during lunch, he took a call from home and was upset. Whether you call it 'upset,' or bothered, frustrated, irritated, annoyed, angry or discouraged, discontinue work when you feel that way and perform a safety stop.

Rerun—Have you ever been driving, traveling on an hour-long drive, for example? You find yourself pulling into the parking lot of your destination only to realize that you don't remember driving the last half hour, your mind was lost in thought. Driving is so routine, that you 'zoned' out, thinking of other, more important things while you were behind the wheel. The same thing happens in our work, from time to time, when we are working on a task that is a rerun; it's routine. When we know a task is a rerun, if the task has a high likelihood of 'zoning,' and/or you find yourself 'zoned' out, stop! Complacency kills, and a rerun is a form of complacency. When this happens, perform a safety stop to get your mind back on track.

Tired—"The disastrous crash of the Challenger Space Shuttle in 1986 in which seven astronauts lost their lives, occurred after NASA officials made an ill faded judgment to go ahead with the launch after having already worked for more than 20 consecutive hours," writes Jim Loehr and Tony Schwartz in the eye-opening book on personal energy called, *The Power of Full Engagement*. "The longer, more continuously, and later at night you work," Loehr and Schwartz continue, "the less efficient and more mistake prone you become." Understanding this simple statement for what it is means; that when tired, it's like we are driving on ice. We know we have to move ahead, but we do so with much more caution. If you find yourself working 'tired' for any number of reasons, stop! Perform a safety stop then move forward.

Once you recognize a H.U.R.T. sign, perform a safety stop immediately! A safety stop is a simple two-minute 'stop.' It is equally effective for an individual or an entire crew. During the safety stop, re-evaluate all surroundings, hazards, and equally important, attitudes. Identify the source of the H.U.R.T. and form a simple plan so that work may continue, safely! Once the H.U.R.T. is gone, the chances of injury or incident decline significantly. Recognizing the H.U.R.T. warning signals before an incident can be the difference between going home safe and without incident, or not. If you don't believe me, just ask me about a cold and rainy night in October!

29 Beyond Zero
How to Send Your People Home in 'Better' Shape than They Came In!

A number of years ago I attended a safety banquet for a utility company. The purpose of the event was to celebrate, with workers and their spouses, the fact that the work group had worked the entire year without a lost workday case. During the event, a vice president took the podium to share a few words. I still remember what he said. "Your families let go of you each morning." He began. "You belong to them and we borrow you for eight hours of work. Your families expect and deserve to have you back at the end of each day, whole and healthy."

That being said, the term zero injuries is becoming a more common trend or theme for industry. Many organizations have 'target zero' or 'zero is possible' posters pasted all over break rooms and on truck bumpers. In fact, the Center for Disease Control and Prevention, located in Atlanta would consider safety improvements as "One of the greatest health achievements in the 20th century." According to the CDC, the workplace today is, on average, nearly 40,000 lives a year safer compared to the 1930s. Yet, it could be argued that safety gains over the last half-decade have flattened. For example, the Bureau of Labor Statistics reported that fatalities were record low in 2003, with 5,575 workplace deaths; that number has slowly risen since the low point. It begs the question, have we reached a backstop called zero? If that is the case, in order to move us into the next level of safety performance, it may force safety professionals and leaders across the country to ask, what's beyond zero. If I can send my people home 'in the same shape' as they arrived, can I return them 'better' than they were when they arrived?

How to Move Beyond Zero

Start with a Firm Foundation—I believe it was Stephen Covey who said, "Systems are designed to produce the results they are currently producing. Organizations are systems. If you don't like the results you are producing—change the system." First and foremost, we can't move beyond zero until we are 'near zero.' If your organization isn't there, stop reading now and return to this article after sustained safety success is established. Near zero is found in safety statistics, but those numbers are driven from the establishment of an effective safety system. These systems make the firm foundation of safety success and include; safety and hazard awareness programs, safety committee processes, accountability systems, senior leadership and engagement, recognition programs, etc. Once these are in place and we have measured success over time, we are ready to look beyond zero at the possibilities that exist.

Stretch and Flex Programs—About a decade and a half ago when I was an overhead electric lineman, I was reporting to a job site in Boonville, MO. Reporting alongside our line crew was a railroad crew. Before the rail crew began work, they were required to stretch and flex. In so doing, they had about as much passion as a boy who was required to kiss his sister! Their heart was not in it! Fast forward to last summer when I spent some time with major construction companies. I was amazed and pleased to find that they had an aggressive stretch and flex program. They have dubbed their employees as 'construction athletes.' And, their workers understood the importance of such programs. In Boonville, as the old timers 'made fun' of the rail crew, they also popped aspirin all day to help aching muscles and joints. Physical work is tough and can take a toll on the body over time.

Gut Check—If we change our perspective to 'beyond zero,' how does that change the view of stretching and flexing over time? How important is proper lifting, body positioning, micro stretch breaks and flex programs? We are only given one body and it's not to be 'spent' at work. Instead, our bodies can be used to earn a living so that we may enjoy life after work. What's beyond zero? I'm not sure, but I do know it begins when our collective work groups can touch their toes!

Energy Breaks—"A man was walking through town," an old story begins, "and notices a friend on a bridge getting ready to jump. He quickly runs over and tells his friend not to jump. Come down, instead, and talk about the problem. So the friend did come down, they talked and two hours later they both jumped!" What is arguably the number one danger in the workplace? Complacency. What is the opposite of an enthusiastic and proactive energy level? Yes, it is complacency and it can hurt and even kill! Here is a simple litmus test on organizational energy. If 'one' is very low energy and 'ten' is optimal energy, rate the following safety activities; your most recent safety meeting, your organization's safety committee engagement, the last safety observation activity, the most recent job briefing, etc. Unfortunately, if we are honest, most organizations are hovering around two or three on the 'ten' scale. All too often, the work environment tends to be an energy hole, draining away personal energy and engagement. Commitment, feedback and compliance might be jumping off the bridge with this low energy.

Gut Check—Would a higher level of energy help workplace safety? What results can you achieve with a higher level of enthusiasm and energy level? What are some tools that could be used to raise an organization's energy levels? Can our people leave a job site physically tired yet emotionally energized and how would that wipe out workplace complacency?

Community—"No one told us to take the fun out of work," an anonymous quote reads, "we did that on our own." In their ground breaking book, *First Break All of the Rules—What the World's Greatest Managers do Differently;* authors Marcus Buckingham and Curt Coffman reached some interesting conclusions. They tackled a mountain of data, over 80,000 manager interviews from across the country and from diverse industries, to determine that community plays a substantial role in employee performance. Among other conclusions, they found the following workplace qualities in high performing employees; an environment where supervisors care for employees as people, development is encouraged, opinions count, praise is given often, there is commitment to quality and employees have a best friend at work. To that end, how is the community in your workplace. Do employees have friends? Does your organization care? What kind of feedback is given and how often? In the end, community is a place where we care and want to be present. What can we do to move our organizations toward community?

Gut Check—Would feedback for safe work rules and safe work compliance be better or worse in a caring environment or in a 'community'? What are five small things your organization could begin doing tomorrow to encourage community?

"Don't be afraid to give up the good," said Kenny Rogers, singer and actor, "to go for the great!" After the safety system foundation is poured and our organizations are 'near' zero it seems to be time to look beyond zero. It seems to be appropriate to look at how we can return our people home 'better' than they came to work. In the end, work has mostly been viewed as a place where one 'gives' life. Why, however, can't we look beyond zero and view work as a place where we actually get life, grow and become a better person? Once strong programs are in place and the safety foundation is poured, if we begin to think beyond zero, amazing thoughts, and results, can happen. These are just a few ideas . . . what are yours?

30 Safety's Broken Windows

Why Wheel Chocks and Steel-Toed Shoes Really Matter!

In the mid 1980s, New York had a crime problem. While they weren't the only major city that was seemingly losing the crime battle, they took a very unique approach in trying to solve it. What did they do to solve a crime epidemic? They started with graffiti on the subway!

In 1985, the New York City Transit Authority was at a crossroads. The system was aging and crime on subways was seemingly out of control. The Transit Authority hired a new director, David Gunn, to solve the problem. Gunn inherited a list of complex issues such as budgets, crime on subways, an aging system, questionable management structure, etc. Gunn would lead a multi-billon dollar effort to rebuild this system and many experts were telling him to start with system reliability and violent crimes. Instead, he ignored the advice of the consultants and decided to try an unusual tact for improvement to turn around this seemingly broken system, graffiti. He set the goal to eliminate graffiti and vandalism from the subway system.

He began, car-by-car and train-by-train to retake the subway system. He began with the Number-7 train that connects Queens to mid-town Manhattan, taking the train out of service and removing all vandalism and graffiti. He established a rule that once a car was cleaned and put back into service any new graffiti would be removed immediately. Cleaning stations were established and as a train finished its run, it would be pulled into the cleaning station and any new graffiti removed. If it could not be removed, the train would be pulled from service until all signs of the vandalism were gone. This effort, which began in 1985, took until 1990.

In 1990, the Transit Authority named William Bratton to head the Transit Authority Police. In this role, Bratton, a military veteran and lifelong law enforcement officer, took another unique position in solving the crime problem. With felonies and violent crimes on the subway system at record levels, Bratton did not target subway drug use or violent offenders. Instead, Bratton, like Gunn, defied experts and set a goal of cracking down on fare beating. You know, fare beaters are those who jump over the turn-styles to save the $1.25 subway fare.

Bratton wanted to put an end to the 170,000 people who were entering the system without paying. He began to set up undercover teams who worked to catch fare beaters. Once nabbed, these fare beaters were cuffed and placed in a holding area within the subway terminal. It was a location that everyone entering and leaving the subway could see. Once the holding cell was full, all detained would be hauled to a booking bus, processed, finger printed and a criminal background check performed. Like Gunn with graffiti, Bratton wanted to send a message to fare beaters that this behavior would no longer be tolerated.

The broken windows theory was articulated by Criminologist James Q. Wilson and George L. Kelling in 1982. Their theory reasoned that if a building has a broken window that goes unrepaired for a long period of time, vandals will recognize the broken window as a signal that no one cares. Since no one cares, vandals will be inclined to break more windows. Broken windows will quickly lead to other forms of vandalism then breaking into the building, which will lead to more crimes both in the building and the surrounding areas.

There has been a strong trend over the last half-decade to crack down on the safety equivalent of violent offenders. Thousands of organizations and industries over the last few years have established 'rules to live by' or 'cardinal sin' lists. These are lists of safety violations so egregious that if violated, one could experience a life changing event and even death. Depending on the industry, this can include failure to wear fall protection or effectively tie off, failure to effectively guard against electrical hazards or improper confined space entry. If employees are 'caught' violating one of these 'felony' type safety rules, they are immediately suspended.

"Don't be afraid to give your best to what seemingly are small jobs," motivator and trainer Dale Carnegie would say, "every time you conquer one it makes you that much stronger. If you do the little jobs well, the big ones tend to take care of themselves." While I don't disagree with cardinal rules, I think it identifies a trend of only looking toward 'the big one.' This is done often at the expense of wheel chocks and earplugs. "The graffiti was symbolic of the collapse of the system," Gunn later said. "When you look at the process of rebuilding the system and moral, you had to win the battle of graffiti." For Gunn, graffiti was a broken window. The idea was to communicate a very clear message to vandals— the windows were no longer broken! For Bratton, the broken window was fare beating. It represented a signal that led to more violent crimes. Once one or two people were fare beating, others would think that if someone else wasn't paying then they weren't going to pay either. If one window is broke, then another one won't hurt.

Keep the focus the 'big picture' and widen that picture to identify safety's broken windows. Where is your graffiti and fare beating? What signal is sent when employees turn the other check on very small and simple rules such as wheel chocks, earplugs or having the right number of buttons buttoned on a flame retardant shirt? These, and other seemingly 'insignificant' safety practices are our broken windows. This is a signal from management to the employees of what is important. In learning from Gunn and Bratton, we understand that a vigilant focus on the single broken window might actually net the results for which we have been looking. In the end, Bratton found that one out of 7 arrested for fare beating had an outstanding warrant. He found that one out of 20 were carrying an illegal weapon. In safety, those who violate the smaller, seemingly less 'dangerous' safety rules are giving a signal of a broken window. A window in need of repair before it's too late and greater damage is done. Erase the graffiti today!

31 Closing the Say/Do Gap

How the Signals We Send Could be Producing the Results We Don't Want!

A number of years ago, I was asked to be an honorary coach of a little league baseball team. For the opening game of the summer season, I would be their motivational leader; I would also have the honor of being third base coach and have the responsibility for drinks and snacks.

In this particular small town, little league baseball is a big deal. On opening day, players from the youngest age group, six and seven, all the way up to the 13 and 14-year-olds parade through Main Street. Riding on trucks and trailers, these players throw candy and wave to family, friends and on-lookers. After the parade, I met the team. Gathering them in a huddle, I introduced myself and asked what was the purpose of playing this game? Most boys said the purpose was to have fun. I reminded them that the purpose of any game is to win! "Today, if you are batting and someone is on base, you look at me. I will give you a series of baseball signs. Based on this sign, you will know if you should bunt, try to walk or swing away." These 9 and 10-year-old boys were looking at me like I was nuts. After a long and serious pause, I smiled and said, "I'm kidding boys, you're right. We are here to have fun. If you want to walk, you walk. If you want to try for a home run, swing away." With a long sigh of relief, we all stacked hands and yelled 'team!'

Now, this has probably never happened in little league baseball but due to my positive pre-game pep talk, we retired the other team in order. When it was our turn to bat, our first hitter walked. In my short time coaching Josh, this next batter, he seemed to be a bright and athletic kid. As he stepped in the batter's box, I flashed some pretend signs just for fun. He waited, watched and nodded that he understood the signs . . . the other team didn't know what to think. Yet, Josh was about to throw the biggest curve ball of all. Right before his first pitch, he yelled 'time out.' The empire yelled 'time,' and Josh came jogging toward me.

Now I'm thinking, Josh, what are you doing, they are just pretend signs but I didn't panic because I watch the supper Nanny. I know just what to do. I get down on Josh's level and say, "Josh, my friend, what are you doing, they are just pretend signs." To which Josh responds, "Yeah, I know coach Forck, I'm just pretending like I didn't read them!"

Getting Out of the Unhealthy Normal!—One of the best examples of an unhealthy normal is when one travels to Beijing, China, for example. On arriving, your eyes are irritated by the smog and pollution. Yet, after several days, it goes unnoticed. You think nothing more about it until you arrive back in the United States, step off the plane and breathe cleaner air; only then do you realize you had accepted the smog as 'just part of life.'

Safety is a game of signals and actions. As leaders in an organization, we are often like the third base coach, offering verbal direction and flashing signals based on how the game is unfolding. The problem is we often find that we are stuck in an unhealthy normal. Levels of our organization don't agree on the game goals, safety. Instead, we are operating in an environment where the verbal direction is inconsistent with the game signals. This is a 'say/do gap.'

The truth about this gap is that our people more often than not, take their signals by what we do, not what we say. They will, however, clearly know there is a difference between what is said and done. It is in this say/do gap that your safety culture is defined. It's in this gap where choices are made that sometimes get people hurt, even killed.

What signals are you sending? To find out what signals you and your organization are sending, ask front line workers what is important. Not what the organizazation says is important but what front line workers feel is important based on the signals of leadership. As safety professionals, it is our responsibility to accurately define what is in this gap and communicate the findings. In addition, we can unknowingly contribute to this gap in many of the processes and procedures. How do we unknowingly contribute to the say/do wedge and how can we fix it? Below are a few examples of common 'say' mistakes and 'do' fixes to consider.

The SAY Mistake—*Quick and ineffective discipline.* This is sometimes a touchy subject so let me be clear; discipline can be a necessary and required tool to change behavior. Yet, all too often, it is used as 'the' problem solver instead of just one piece to the cultural puzzle. For example, an employee will receive signals to act in a certain fashion, even if the action cuts a small corner off the safety rulebook. After dozens, if not hundreds of repeated actions, an incident will occur as a result of the action. After an incident analysis, the primary finding will be 'worker violated safe work rules,' and discipline will result . . . case closed.

The DO Fix—*Understand that the actions of employees are part of a larger system, often referred to as the culture.* Steven Simon, in an article entitled, "Transforming Safety Culture," April 2009 edition of *Professional Safety* stated, "Safety excellence is a product not only of the right programs, but also of the right culture." The mistake from the above example is that many organizations discount or simply fail to investigate the 'signals' the employee received that led to the incident. It's in this much more painful and delicate investigative process that unhealthy normals are found and say/do gaps are discovered. Without this step, the above-mentioned case is yet another signal to the organization of what is really important.

The SAY Mistake—*The rush to PPE.* As safety professionals we know and understand the hazard management principles, engineer, administrate, PPE. As you know, new hazards are discovered and presented to management. These new hazards might be discovered due to the purchase of new equipment, an added process or perhaps from a new, higher level of awareness. The solution to these problems is all too often the rush to PPE!

The DO Fix—*Eliminating the hazard*. Willie Hammer, in his insightful book, *Occupational Safety Management and Engineering, Fourth Edition*, wrote, "The most effective method of avoiding accidents is with designs that are 'intrinsically safe.' Intrinsic safety can be achieved by either of two methods: (1) eliminating the hazard entirely, or (2) limiting the hazard to a level below which it can do no harm. Under either condition there will be no possible accident resulting from the hazard in question." The most effective way to achieve this is in the 'design' or engineering phase. Yet, all too often when dealing with a hearing conservation issue, for example, earplugs will be the quick decision with a noise reduction opportunity at the source going without mention. Or, when sharp edges are produced in a manufacturing process, gloves will be the default response, versus analyzing and solving the problem at the source. When there is a rush to PPE, what signal does that send to the work force? What hazards does this leave on the job? What gap does this create between say and do?

The SAY Mistake—*The 're' solution*. One of my favorite quotes is from an unknown author and reads, "Life turns out best for those who make the best of how life turns out." Unfortunately, injuries happen and when they do, they present an opportunity to learn and prevent reoccurrence; in other words, "to make the best out of the way things turned out." Unfortunately, many organizations fail to find the root cause and recommend a 're' solution. 'Re' solutions include, re-training, re-writing the procedure, re-issue the safety rule or policy, re-view in a safety meeting, etc. While 're' solutions are sometimes part of the recommendation package, they will not prevent re-occurrence!

The DO Fix—*Elimination*. When it comes to incidents, we understand that severity is simply a matter of luck. And, we understand that if we don't dig deep enough to find the root cause, we are accepting the fact that the incident and injury will happen again, and again and again. The only thing that will change will be the person who is hurt and the severity of the injury. Given this fact, we have to be dedicated to finding a root cause in each and every incident. Not only do we find the root cause, we must get out of the 're' rut and follow through on corrective actions that will eliminate the probability of reoccurrence. "Difficulty," Edward Murrow once said, "is the excuse history never accepts." What do our organizations accept in terms of injuries, root cause solutions and space in the say/do gap?

The SAY Mistake—*It's about them.* In truth, it is the physical workers that have the most exposure and they should be the most 'jazzed' about their safety and the safety program of the organization. Yet many times, there is enough enthusiasm for a sleeping contest. When the energy level is low and the sense of ownership by workers seems non-existent, it is often the knee-jerk reaction for management to walk away. To place the ownership where it belongs. It's about them and they don't seem to care so I don't care! When you have this feeling remember, they are only reacting to signals!

The DO Fix—*It's management's to lead!* There is arguably no better reference on a management involvement than Dan Peterson's classic book entitled, *Safety by Objectives.* There he writes, "If properly installed and well done, there should be no problem keeping it going." Peterson is referring to the management system of involvement centered around specific tasks and responsibilities of management, at all levels. When workers withdraw and management wants to disengage, what signal is being sent? What systems are in place? Finally, what is filling the say/do gap?

In short, every organization has a say/do gap. It's there, however wide or narrow, and may even be contributing to an 'unhealthy' normal within an organization. Find that gap today. Look deep, to find what's really in that space. When we work to narrow that gap, it will mean the signals the organization sends and the words the organization says align; and that's one powerful and safe culture.

32 Leaving the C.A.V.E.
Three Thoughts on Dealing with Difficult People

The C.A.V.E. People

C.A.V.E. stands for Citizen Against Virtually Everything and I would bet that when you first read the word C.A.V.E. people not only did you smile because you have C.A.V.E. people in your organization; you were probably able to picture at least one or two specific people! Each organization has C.A.V.E. people and since they drain energy and time from organizations, we struggle with how to effectively manage this category. I apply the following logic to C.A.V.E. people management.

Consider this; if you had the opportunity to work with one member of my team to increase performance by fifty percent, who would you chose? The natural tendency is to pick a low performer, a C.A.V.E. person. We believe that we can help this person, change this person and make this person a better performer. But, the reality is, I would need to look at the numbers. Would I assist a person that produces two units a month, increasing his output would mean I get three units after my intervention. Or, would I target a performer who produces 10 units per month. Increasing her output means I would yield 15 units after the intervention! I would, of course, work with the 'better' performers. The point is simple, don't get caught in the C.A.V.E. people's trap. Organizations spend valuable energy, training, and time trying to 'reform' C.A.V.E. people. Stop! Instead, invest in both the senior leaders and the Rank and File, because that is where you will move from 10 to 15 units. That being said, C.A.V.E. people can't live rent free. The following are thoughts to consider in a procedure for the C.A.V.E. people.

We Can't Change Them!—Business researchers Marcus Buckingham and

Curt Coffman unearthed some ground-breaking discoveries in their book, *Break All the Rules: What the World's Greatest Managers Do Differently*. The pair painstakingly read through nearly 80,000 interviews with managers conducted by the Gallop polling agency. They were looking for the best-kept secrets of great managers. They wanted to debunk myths and share lessons learned by the best of the best. Among other traits, Buckingham and Coffman found that great managers understand that they can't change people; they are who they are! For C.A.V.E. people, it means that we need to stop trying to change them. The point is, don't invest in this group, instead, tolerate this group and hold them accountable. It's important that we do not mandate extra training or resources to this group, because it won't return our investment. Do what great managers do, and deal with what you have.

Avoid Avoidance—Mark Towers, in his fun and eye-opening paper entitled, *How to Deal with C.A.V.E. People: Citizens Against Virtually Everything* wrote, "Don't let C.A.V.E. people live "rent free." The C.A.V.E. person oval of the D.O.O.L. model can be somewhat difficult and the natural tendency is to avoid them . . . let them exist 'rent free.' Instead, charge rent and go an extra step to identify specific people who are in this group. After identifying who they are, make sure that they are receiving and participating in all mandated training, job observations, safety meetings, procedures, etc. in which everyone else is participating. Charge rent and hold C.A.V.E. feet over the fire to perform; just like everyone else.

Finally, Quarantine to Save the Ship!—Dr. Lynne Offerman, a professor of organizational studies, in the *Harvard Business Review*, wrote the following about the explorer Ernest Shackleton;

"He was stranded on an ice pack crossing 800 miles of stormy seas in an open boat; Shackleton knew the deadly consequences of dissension and therefore focused his attention on preserving his team's unity. He was happy to delegate many essential tasks to subordinates, even putting one man in charge of 22 others at a camp while he sailed off with the remainder of the crew to get assistance. But the one task he reserved for himself was the management of the malcontents, whom he kept close by at all times. Amazingly, the entire crew survived the more than 15-month ordeal in fairly good health, and eight members even joined Shackleton on a subsequent expedition."

Instead of assigning the organization's C.A.V.E. people to a number of different supervisors and managers, to divide the pain, group them together, as appropriate, under the watchful eyes of a capable leader(s). The role here is to make sure they are paying rent and minimizing their potential ability to pull energy from the organization.

"It's not the hand we are dealt," an old anonymous saying reads, "but how we play the hand that makes all of the difference." In work, we will not escape the chilling effects of C.A.V.E. people but we can minimize these effects if we play our cards right.

33 The Safety Committee MAP (Monthly Action Plan)

Your Course to Safety Committee Success

To create real change within an organization, one must alter resources, values or procedures, this according to Clayton M. Christensen author of the *New York Times* bestseller, *An Innovator's Solution*. To be successful, safety committees need the time and budget (resources). The organization must establish a value base that fosters safety. But, once those are in place, we still need a means (procedure) to cultivate safety committee success.

Finding the Pot Holes . . .

In the past, safety committees have done as well as can be expected. We have asked them to set goals in an attempt to ensure their proactive success and we have encouraged and supported these goals. But in failing to offer a specific action plan we have asked our committees to 'play ball' without defining the field. Imagine a baseball team attempting to 'play' without a defined field, a pitcher's mound, home plate, batter's box, bases, foul lines or assigned positions. The team can play hard but without a defined field it is hard to win.

This is similar to what our committees have been doing. Instead of being consistently proactive and awareness driven, they become reactive, engaged in few if any safety awareness activities. Most exhaust a great deal of time discussing items that are not safety related. Few have a plan for the next month. Many, in an attempt to do something, grab issues that should have already been addressed by management. And, management lets the committee have the issue, thankful that they are doing something. They are playing hard, they are just not sure of the 'field.'

A wise truth reads, "The arrow that hits the bull's eye is the result of a hundred misses." Our committees have shot many arrows, some hitting but most missing. Over the last few years, I have taken keen note of what arrows have hit and which ones missed. In so doing, I introduce the Safety Committee MAP (<u>M</u>onthly <u>A</u>wareness <u>P</u>lan), a strategy that will hit the bull's-eye month after month.

Begin the Drive . . .

To begin the MAP process, brainstorm using the following questions. Where are our workers taking shortcuts? What are we doing that gets us hurt? What safety rules are neglected? Which safety rules are the most important? What does our previous peer observation data tell us? What training do we need this year? What needs does your community have? What holidays do you celebrate?

Twenty minutes of brainstorming can produce a hundred or more possible themes. Pick your favorite twelve. These will be your safety themes for the year (one per month). You and the committee have the discretion to change a theme if needed but at least you have established your framework for the year.

Shifting into Gear . . .

For each theme the committee should develop a MAP, each MAP should include the following:

Two Safety Activities—A safety activity is something done outside of a safety meeting that brings awareness to safety and your monthly theme. It might be a note taped to a locker, stickers or awareness items in trucks or at workstations, etc.

An Involved Safety Meeting Activity—Is a safety meeting that involves the entire group. It can be a game, activity or exercise that gets everyone involved.

There are references that can help with this such as my book, *ISMA—Involved Safety Meeting Activities: 101 Ways to Get Your People Involved.*

An Outside Safety Speaker—This is an individual from outside of the work group that can share his/her insights on the monthly theme. It might be a subject matter expert from the community (police officer, doctor) or an individual from another part of the company.

Distribute at Least One Type of Safety Educational Material During the Month—This is an item such as a quiz, crossword puzzle or other educational game. It can be used during a safety meeting to educate the group on the monthly theme. These can be easily found on the Internet or created using inexpensive 'word game' software.

Finally, Hand Out at Least One Safety Trinket—A safety trinket is a tangible item that is given to employees to bring continued awareness to the monthly safety theme. For example, a Crunch candy bar can be used to remind co-workers not to be caught in a 'line of fire' crunch.

Once each activity, speaker, involved safety meeting, etc. is chosen assign responsibility for that activity to members of the committee. This ensures the task is completed on time.

Why Unfold the MAP?

The MAP concept does seven unbelievable things for your organization.

The MAP Shows Caring—The soft side of safety is about relationships. Our people don't care what we know until they know we care. People know that you care when you are there for them, out front and visible. The MAP framework fosters caring and relationships.

Involvement—So many committees have asked me in the past, how do I get my co-workers involved? We typically think someone is involvement when they come forward to deliver a safety meeting or write a safety quiz. Let's face it, that doesn't happen often. When following the MAP, everyone is involved. When I hand an employee a safety trinket or include him/her in an involved safety meeting activity he/she just became involved in safety. And, involvement equals participation, which leads to ownership that equals results!

The MAP Fosters Safety Committee Teamwork—Did you know that a Belgium Draft Horse could pull 8,000 pounds by himself? If you team two horses together they can pull 18,000 pounds. If you train them to work as a team, they can pull 25,000 pounds. Too often, committees are led by one or two individuals. The MAP engages the entire committee, in the end, making them more effective.

The MAP Keeps Focus—Committees will be focused on their monthly activities and are more likely to divert concerns that can be handled outside of the committee to the proper parties. You can't hit a target you can't see but with this plan you've defined your targets and your arrows, it's a win-win.

The MAP is High Safety Energy Awareness—'Reaction' and 'creation' have the same eight letters. When you react you respond in the same fashion in which you have always responded. Yet when you create, you think before you act. Maintaining high-energy safety awareness makes our workers think, thus creating safe situations instead of reacting to old habits.

The MAP Communicates to the Work Group—When the committee is being reactive, the work group often doesn't even know they exist; the committee is contributing nothing to their work group. When following the MAP framework there will be no doubt the committee is out in front and that workers are actively supporting the safety of their peers.

The MAP Changes Behaviors in a Positive Direction—Since the monthly themes are picked based on a high number of incidents in a certain area, past injuries or weak safety habits, these monthly activities continue to encourage safe behavior, challenging and changing at-risk behavior.

Driving Home Safe . . .

Have you ever wondered why the white strip is painted on either edge of a narrow two-lane highway? Well, when oncoming headlights, fog, snow or rain blind a driver, all he needs to do is find and focus on the white line. If the driver doesn't have a focus, he can easily lose control and drive off of the road. If he focuses on the yellow centerline, he can easily cross that line into oncoming traffic. Although visibility is poor and the driver can't see ahead, he knows that continued long-term focus on the white line will get him to his destination safely. The Safety Committee MAP (Monthly Action Plan) is our 'saving' white strip. Open the MAP and drive knowing that you are changing your safety culture, one MAP at a time.

Safety Committees are a tremendous investment in time and money for organizations yet most of the time, the return on the investment falls short. Safety Committees fail to function at a high level because they simply don't know how. Following the Safety Committee MAP (Monthly Action Plan) offers a simple easy to use plan to ensure Safety Committee Success, month after month.

34 What's Beyond Behavior-Based Safety?

In the last four decades nothing has changed the face of safety more than the establishment of OSHA and Behavior-Based Safety (BBS) Programs. In fact, you probably can't find a company today that doesn't have some elements of BBS within their current safety process. That being said, one could argue that BBS is nothing new and really started in the 1930s with William Heinrich and Traveler's Insurance. As you may know, Heinrich was responsible for reviewing injury and incident reports submitted by supervisors. Over time, Heinrich began to realize that most injuries were 'man-failures'. After much analysis, Heinrich published his findings, which stated that 90% of all injuries and incidents were caused by these 'man-failures' or what was later termed, unsafe acts. In its simplest form, Behavior-Based Safety could be defined as the programs, systems and resources dedicated to identifying and eliminating these unsafe acts or human errors.

Today, organizations across the country have the same problem; safety results are flat. There seems to be a growing urgency to act, but many organizations are not sure exactly what to do. It seems the time has come to ask the question, what is beyond Behavior-Based Safety? Below are a few thoughts on how we can move some traditional behavior-based safety programs into the 21st century.

Leadership to the Right Level—Behavior-based safety has done a great job of putting a focus on organizational leadership. In the last few years, one couldn't throw a dart at a safety magazine without hitting an article on leadership. For the most part, these articles focus on getting managers, vice presidents and CEOs engaged in the process. And, that is a very important part of the leadership equation, but it is leaving out the part of the equation with the most influence over our worker's choices.

John Maxwell, an author and noted authority on leadership defines a leader, "as one with influence over others." In our work environments, we have rank and file employees who have great 'influence over his or her co-workers.' These informal leaders are subject matter experts, those who excel in their trade, sometimes they are veteran employees or simply charismatic leaders. If we are to help our organizations reach a new level of safety success, we will have to engage senior leaders, but we must also recognize, engage and foster these rank and file leaders, our informal safety leaders (ISLs). Senior leaders and informal leaders working together can achieve great success.

Substitute 'Get Up and Do' for 'Sit and Listen'—Safety Meetings are a key part of eliminating unsafe acts. The National Safety Council has issued data on adult learning and retention. In the end, if our employees can put their hands on it and feel it, in some cases their retention will be nearly four times greater than simply sitting and listening to the same material. So, what's the problem. The problem is that most organizations are stuck on sit and listen safety meetings— meetings that lack energy and educational value. If we are going to move beyond Behavior-Based Safety, we will have to shift to Involved Safety Meeting Activities, or ISMAs. The formula is simple, when preparing for safety meetings, ask some key questions, how can I bring in the tools and equipment to the meeting, or the meeting to the tools and equipment to get my workers involved, instead of reading or showing a picture? Can I break meeting participants into smaller groups for an activity? What activity can I do to start the meeting, such as a quiz or quick game, in order to immediately engage participants? What motivational activities can I employ to build awareness? Answering these key questions is the lever to pull that will put you on the road to more active safety meetings and results. (Note; check out my book, *ISMA—Involved Safety Meeting Activities: 101 Ways to Get Your People Involved* as a resource).

Move from Rules to Relationships—Randy was a tough, hard to get to know, son of a gun. He had more than twenty-five years of experience and didn't get along the best with the local management. I knew Randy and had actually worked with him when I was an apprentice. I was now the safety supervisor for the area. In this roll I was responsible for nearly 400 linemen, substation workers and gas employees in rural Missouri. Knowing that I couldn't see everyone in the course of a month, I decided to show each I cared in a different way. I got with Human Resources and received a list of everyone's birthday. Then, I wrote each person a birthday note. For Randy I said something like, "Happy Birthday, I really liked working with you back in the day. I always liked your funny stories. Work safe, Matt." I sent it and forgot about it. About five years later I happened to be in Randy's show-up location and there on his locker with the pictures of his wife and kids was a faded piece of paper. I recognized it immediately; it was my note to him.

Family counselor and author Josh McDowell wrote, "Rules without relationship causes rebellion." If I needed to talk safety or enforce a rule with Randy, I'm confident he would have listened because he knew that I cared. Changing behavior is first about creating a relationship . . . that is the next step in Behavior-Based Safety.

Reward Recognition and Appreciation, Not a Rewards Program— "Recognition," says Richard M. Kovacevich CEO of Wells Fargo, "is American's most underused motivational tool!" One commonly used Behavior-Based Safety element is an employee awards program; generally giving a small gift for performing certain safety activities over time. Unfortunately, these programs have turned to a 'check the box and get a prize' system. In the insightful book entitled, *The Invisible Employee* one can find remarkable statistics on recognition and appreciation. "According to a 2003 survey, 90% of workers say they want their leaders to notice their efforts and improve their recognition and rewards." In addition, "In an ongoing Gallup survey of more than 4 million employees worldwide, there is remarkable evidence of the business impact of recognition and praise. In a supporting analysis of 10,000 business units within 30 industries, Gallup found that employees who are recognized regularly increase their individual productivity, increase engagement among their colleagues, are more likely to stay with their organization, receive higher loyalty and satisfaction scores from customers and have better safety records and fewer accidents on the job." Safety is a tailor made vehicle for appreciation and recognition. The only question is, can we leave our check the box systems for genuine recognition and appreciation? If yes, we can leave our current results and move into results we never thought possible.

Downstream to Upstream in Real Time—Behavior-Based Safety has encouraged safety measurements to be more than a simple review of the injury roster. Although recordable and lost time injuries are one measurement of safety success, they are not the only measure. The problem, as you know, with looking at injury data is that it is downstream, or reactive. BBS has encouraged organizations to look upstream at proactive data like unsafe acts from safety observations, safety audit scores, inspection results, etc. The problem with this proactive data is at least two-fold. First, it is not widely published throughout the organization, like injury statistics are; and once this data arrives on the desk of senior leaders, it is old and outdated. Moving beyond BBS means we use today's technology to make proactive data real time. It hits the desk of safety staff and senior leaders as it happens and is recorded. This way, senior leaders can offer the right recognition for a job well done or quickly intervene, when needed.

In the end BBS, along with OSHA, were two key drivers in leading us to our current level of safety results. Today it is time to ask the question, what is beyond BBS . . . and how can we lead our organizations there?

35 Making Safety Top of Mind

Creating Awareness to Reduce Injuries and Near Miss Incidents

What do your workers think of first when they strap on a harness, put on a hard hat or head to the floor? Chances are it's not safety! You see, for the most part, our people are trained, skilled and knowledgeable. They know what to do and when to do it. Because of this, putting on safety equipment, negotiating high risk environments and managing hazards is 'just part of a day's work.' That is . . . until it's not! We know all too well that conditions, weather, materials, exposures, job sites and surroundings change. What one of our workers anticipates, expects or takes for granted is suddenly not there, which leads to a near miss, or worse. Safety must be 'top of mind' and it's achieved, in part, through awareness. Let me explain.

Willie Hammer, in his instructive and educational book, *Occupational Safety Management and Engineering*, Fourth Edition, writes, "Intermittent safety efforts are generally ineffective. It is necessary to maintain an almost continual program of keeping personnel alert to safety practices." He continues by citing examples of awareness opportunities, "Small folders or booklets available from safety organizations, insurance companies, and the federal government can be given to each employee as he or she enters the plant or at other appropriate times. Slips with printed safety messages can be added to pay envelopes or attached to pay checks. Place mats or napkins with interesting messages on safety and accident prevention." Top of mind wasn't a catch phrase when Mr. Hammer wrote this but if it was, I think he would have said, "Safety needs to be top of mind, and that's achieved through awareness!"

To that end, I wanted to introduce you to Joe. Joe was a first line supervisor for a mid-west utility. He had been a line worker for nearly 20 years and had worked almost another 20 as a first line supervisor. He didn't read Willie Hammer nor had he ever heard the term, "top of mind." What Joe did understand however was safety awareness, in other words, how to keep safety top of mind. Joe was always putting together a contest, safety breakfast, hard hat sticker, slogan or skit. He would stop at nothing to make safety fresh and memorable. He wanted his people to be aware which in his opinion was a key factor in safe choices. I haven't worked with Joe for over a decade but I still remember a number of his favorite safety awareness techniques and theories—and now you can use them too to make safety for your people top of mind!

Energy—Joe realized that energy was key, and he was right. Jim Loehr and Tony Schwartz, in their book, *The Power of Full Engagement: Managing Energy, Not Time, is the Key to High Performance and Personal Renewal,* wrote, "Leaders are the stewards of organizational energy!" How are we managing our organization's energy?

Turn it Upside Down—One of Joe's favorite tricks was to hang a poster and two weeks later turn it upside down. He always said that after it's there for two weeks, the guys won't see it anymore so it must be turned upside down to be noticed. What else can we turn upside down to make more effective?

Fun—In Joe's awareness world, there was never an end to contests, friendly competitions and wagers. In order to create a buzz, he would purposely make a bet in front of the work group, one that he knew he would lose, so that he would have to pay in front of everyone. It was always high stakes, a Snickers and a Pepsi. And, it was well worth the money that Joe paid in the return of safety awareness.

Food—"The family that eats together stays together." And, the work group that eats together works safe together. Make food a staple of your safety awareness diet.

Visual Reminders—When I think of NASCAR, I think of labels. NASCAR was the first to understand they could sell space on a car, uniform, helmet and visor. But before NASCAR discovered it, Joe was pasting safety reminders on hard hats, job folders and steering wheels. What can you label, sticker or mark? Probably more than you can imagine.

Volunteer for Safety—This safety awareness isn't easy and Joe put together an informal army of volunteers. These people were willing to help and share ideas. Remember, involvement is a naturel reinforce to 'top of mind.'

Take a Test—For most of the hazards and themes Joe would put together a short quiz or word search. It's funny how grown men and women would welcome the challenge to pass a quiz or test in their given craft. How can you use fun tests or quizzes.

Safety Share—After Joe and his team had been at this safety awareness for some time, others began to notice. To that end, his group became a formal mentor for a number of other safety teams. They would share meeting notes, minutes and awareness ideas. It made everyone better.

Set Up the System—I met Joe after he had been in the safety awareness game for nearly a decade. When I met him, awareness was a 'well oiled' system. At the first of the month, Joe, his safety committee and volunteers would capture a theme, one of the major hazards in his work group or an area of recent concern. They would then make a formal plan that included a quiz, stickers, posters, safety awareness items and a contest. Each month . . . that was the system.

Measure It—Joe was very keen on measuring awareness or top of mind. He was always looking and questioning the safety awareness of this work group. At times, he pulled back a little, other times he pressed on the accelerator. The point, he had his finger on the safety awareness pulse for his people. Who has his/her finger on the awareness pulse of your group?

In the end, if we are going to prevent near miss events and injuries, safety must be top of mind. That is a 24/7/365 type of commitment. It means that you are focused, consistent and systematic. If you don't believe me, then take the word of someone with more than two decades of experience and a number of good ideas.

36 Safety Meetings
12 Tips to Blow Them Away and Leave Them Wanting More!

Safety meetings are a big deal. And, I'd bet a large sum of money that your employees are not blown away and do not want more . . . unless they are catching up on more sleep, that is. Safety meetings are 'big' for a number of reasons. They are an important part of our safety program, allowing for instructional and refresher time on important subjects. They are vital for safety awareness, to keep immediate hazards on 'top of mind.' And, safety meetings are a large investment in time and dollars.

Let's dive deeper into the cost equation. If your employees make $20 per hour (many trades and skilled craft are twice this amount) and the employees benefit package is another $20 per hour. Then it costs you $20 per employee for a half hour safety meeting. But wait, production, tools, materials and overhead are often at least a 5:1 ratio of salary. So, when adding that, it costs you $100 per employee for every half hour safety meeting! But wait one more minute, Louis J. DiBerardinis in his book, *Handbook of Occupational Safety and Health* writes, "The average direct and indirect cost of an employee injury accident is $20,000. This translates into the company having to generate $200,000 in sales/services at a 10% profit margin, or $400,000 at a 5% profit margin, to pay for a $20,000 accident." In most cases, this cost can be more than a country club membership!

So, if safety meetings are so important why don't we hit them 'out of the park' each week? That's a good question, why don't we? Below are 12 tips that can help you prepare and execute blow-out safety meetings. These tips will help you raise awareness, effectiveness and results while leaving your people wanting more.

Ask an Injured—Chances are if you look back a decade, you have an employee that has suffered a major injury. Chances are that this injury has a hidden story. Talk to your employee about sharing his/her story, the hidden one focusing mostly on the family, and the impact of this injury at home.

Near Miss Testimony—Much like the 'ask the injured,' we have near misses each day. While it's good for your safety professional or supervisor to share these, it is much better if you can coach those involved in the near miss to share the story.

Involved Safety Meetings—Statistics show that if we read a report or procedure to our workers they will remember about 15% of it. If we show a video, the retention rate increases to about 25%. But, if we involve them so they put their hands on it and get involved, retention rates are near 75%! To that end, exchange sit-and-listen meetings for get-up-and-do meetings. There are a number of resources on the market to learn activities, which can serve as lead-ins to the meeting material. Or, bring tools, equipment and materials into the meetings, so your people can build it or practice it, instead of just sit and hear about it.

Go to the Field—A number of years ago, as an electrical lineworker for a utility, I was asked to plan a safety meeting. The group was young and didn't have much experience with three phase meter change outs. I went to a veteran troubleman and we found a meter that needed to be changed. We contacted the customer and on the appropriate Monday morning, the entire electrical department went to the location, and the veteran troubleman changed out the meter, instructing us along the way. We don't have to stay in the meeting room . . . get out and go to the field!

Ask the Old Guys—In most trades and crafts, the 'old guys' are respected for their knowledge and experience. To that end, put that experience and knowledge to work by asking them to lead a series of safety meetings—they have a lot to teach.

Bring Food—The old saying, 'the family that eats together stays together' seems true. And, 'the work group that eats together is safe together' also has some truth to it. Bring food, and see the excitement grow.

And the Winner Is—Decades ago Henry Ford realized that one shift in a particular factory was out producing others. So, one day he took chalk and scribbled on the floor the number of cars produced by this shift, so that the next shift could see it. That next shift stepped up and out produced the other, and then back and forth they went, both shifts producing more. The point is simple, contests motivate people. I'm not advocating a contest like Henry Ford's since that might lead to shortcuts in the field, but I do strongly support contests in safety meetings that motivate our workers to get involved and have fun.

Ask the CEO—When was the last time the CEO spoke at your safety meeting? When was the last time he was asked? Enough said.

Ask an Outside Speaker—Several years ago when I was working as a safety professional for a utility company, I served most of out-state Missouri. Within my territory was one work location in a town with the population of just over 60,000. The supervisor, who organized the safety meetings, had a goal to bring in one outside speaker per month. He defined an outside speaker as someone outside of the immediate work group. The speaker could be from another part of the company or from outside the company altogether. I first thought this was a lofty goal, one that couldn't be met. Yet, year after year, I saw outside speakers come to safety meetings, one per month. I saw conservation agents, doctors, police chiefs, local coaches, eagle scouts, OSHA reps., managers from other plants, safety professionals from other companies, industry experts, etc. What I really saw were results!

Hire a Speaker—Okay, this seems very self-serving, since I speak to groups for a living, but even if you don't hire me, it seems to make sense to treat your work group once per year to a professional keynote speaker. The energy created by this speaker can generate results for weeks and months to come.

Start the Meeting Right—How you begin a meeting can have great impact on the energy throughout the meeting. To that end, purchase a meeting starter book and use some of the ideas.

Give Something Away—In the world of car sales, Joe Girard is like Michael Jordan. He holds the world record for the most sales and he is quick to remind everyone that these weren't fleet sales but individual sales. His sales were astounding. His monthly retail mark was typically more than an entire dealership, 174 cars and he holds the world record for most career new car sales, 13,001. So, how did he do it, he gave something away. He explains in his book, *How to Sell Anything to Anybody*, that he had desk drawers full of items. Everyone who entered his office received a small token. He had a drawer of stuff for women, one for men and a third for children. When people entered his office, the equation was unbalanced. In the customer's mind, Joe was going to be taking from them through a sale. In order to balance the equation, put the customer at ease, build trust and a relationship, Joe would give each customer an item from his desk drawer, each customer, each time, no exceptions. Give something away. Tie it to your hazards, such as a Crunch candy bar to remind your workers about line of fire 'crunch' hazards. Will this work, well just ask Joe.

In the end, you can blow your workers away with safety meetings and leave them coming back for more. The question is, will you? It is that important . . . good luck and get started!

37 Pulling a Lever
What to do When Nothing Else Works . . .

My grandfather, Otto, was a plumber. He was actually a very good plumber and if one needed plumbing work inside or outside their home in the 1950s or 1960 in or around Jefferson City, Missouri; they called my granddad. The man knew his shi* . . . well you get the idea.

My dad loves to relive the days when he was a kid and helped out in the plumbing business. One of his favorite stories begins early one morning when Otto, my dad and my dad's twin brother, both about twenty years old, were sitting around the breakfast table when the phone rang. Dad's brother answered the phone and the women's voice on the other end was frantic. He couldn't understand a single word; she was hysterical. Handing the phone to Otto, he was able to gather two pieces of information; there was a significant water leak and an address. Hanging up the phone, Otto asked if one of the boys wanted to go along, my father volunteered.

The location was near the downtown area, a mere five-minute truck ride. Arriving on site, they entered the main level of a two-story brick building. The first floor was a doctor's office. There they found a nurse mopping water as quickly as she could. Looking above her head, water was gushing from the ceiling above. Otto quickly told dad to go to the basement and turn off the water to the entire building as Otto headed up stairs. Dad dashed to the basement but found the door to the water valves locked. He ran up the stairs and was just six steps behind as Otto entered the upstairs apartment. There, a man in a suit and tie was on hands and knees trying to hold back a wall of water with every towel he had. In the kitchen, a woman sat crying at the table. Just a few steps away from her was the kitchen sink. Water sprang from the faucet; the nozzle had apparently been blown off. Nearly ankle deep in water, Otto looked at the woman and the faucet then calmly leaned over and pulled the faucet lever, stopping the water.

Once the water was turned off, Otto and my dad talked to the women. They learned that she had turned on the faucet to fill her coffee pot for the morning. When she did, the nozzle blew off spraying water everywhere. The women panicked, trying to stop the flow with towels, cups and anything else she could find. When that didn't work, she froze at the table.

As safety professionals, organizational leaders, supervisors and managers, we are responsible to 'stop leaks.' These 'leaks' come in all shapes and sizes, from people, to procedures, to products and tools—we are responsible. Yet, every once in a while we encounter a leak much like this women. It comes out of nowhere to surprise us and after we have exhausted all of our resources to stop it, it keeps gushing. We often become immobilized, unable to think and act and unable to stop it. When that happens, we must be willing to pull a new lever, take action, to stop the leak once and for all. What follows are three common leaks that often 'freeze' management along with three new levers, ideas, to pull.

Leak: *A Lingering Unsafe Condition*—Probably all of us have been involved with one or two issues where the hazard just couldn't be totally eliminated. Despite your best efforts and efforts of a team, you just could not seem to come up with the right solution or employee buy-in to solve a lingering unsafe condition.

Lever: *Turn the Problem Over to Your Informal Safety Leaders, ISLs*—John Maxwell, a noted authority on leadership defines a leader, "as one with influence over others." In our work environments, we have rank and file employees who have great 'influence over his or her co-workers.' These informal leaders are subject matter experts, those who excel in their trade, sometimes they are veteran employees or simply charismatic leaders. If we can recognize these leaders then engage them and foster their growth over time; we have pulled a lever that can solve nearly any safety problem we have.

A few years ago, a utility was having problems solving fall protection issues within substations. With each substation being different, the problem was two-fold, a set of certified attached points could not be found for each location and many times the only tie point option was at ones feet, making the fall distance too great. Management was frozen after a couple of years of seemingly no results, so a new lever was pulled and the problem given to a group of informal safety leaders. The result was a solid and consistent practice that adhered to standards and workers were willing to buy in.

Leak: *Lack of Energy in Safety Meetings*—How many safety meetings have you participated in that felt more like a funeral than a celebration of safety? And, if workers can't feel energized in a meeting about their safety, how can they be energized about their safety in the field?

Lever: *Involved Safety Meeting Activates*—How much is a safety meeting costing, per person per meeting? Well, if your employees make $20 per hour (many trades and skilled craft are twice this amount) and the employees benefit package is another $20 per hour. Then it costs you $20 per employee for a half hour safety meeting. But wait, production, tools, materials and overhead are often at least a 5:1 ratio of salary. So, when adding that, it costs you $100 per employee for every half hour safety meeting! But wait one more minute, Louis J. DiBerardinis in his book, *Handbook of Occupational Safety and Health* writes, "The average direct and indirect cost of an employee injury accident is $20,000. This translates into the company having to generate $200,000 in sales/services at a 10% profit margin, or $400,000 at a 5% profit margin, to pay for a $20,000 accident." Safety meetings are an important tool to prevent injuries and increase communication and awareness around hazards, and it takes energy to get that done.

The National Safety Council has issued data on adult learning and retention. In the end, if our employees can put their hands on it and feel it, in some cases their retention will be nearly four times greater than simply sitting and listening to the same material. So, when preparing for safety meetings, asks some key questions, how can I bring in the tools and equipment to the meeting, or the meeting to the tools and equipment to get my workers involved, instead of reading or showing a picture? Can I break meeting participants into smaller groups for an activity? What activity can I do to start the meeting, such as a quiz or quick game, in order to immediately engage participants? What motivational activities can I employ to build awareness? Answering these key questions is the lever to pull that will put you on the road to more active safety meetings and results. (Note; check out my book, *ISMA—Involved Safety Meeting Activities: 101 Ways to Get Your People Involved* as a resource).

Leak: *Safety Committee Under-Performing*—It's a classic problem. We know how important safety committees can be to our safety process yet, our committee is under performing. The committee seems to be without direction, goals and purpose.

Lever: *Employ a Monthly Action Plan (MAP) for Long-Term Direction and Success*—Among other duties, a safety committee should be laser-focused on hazard elimination and safety awareness. To achieve best results, employ the safety committee MAP process. It's easy, the committee begins by brainstorming the biggest hazards associated with their work environment. They do this by asking key questions. What has caused injuries in the past? What are the biggest hazards in our workplace? Where are we getting hurt? Where does safety awareness seem to be lacking? This process should result in dozens of thoughts on the white board. Narrow this list to 12, one for each month of the year, then assign each theme to a month.

Once you have your themes aligned to a month, then prepare the following for each month: two Safety Activities, (A safety activity is something done outside of a safety meeting that brings awareness to safety and your monthly theme. It might be a note taped to a locker, stickers or awareness items in trucks or at workstations etc.) an Involved Safety Meeting Activity, (discussed above) an outside safety speaker (This is an individual from outside of the work group that can share his/her insights on the monthly theme. It might be a subject matter expert from the community such as a police officer or doctor or an individual from another part of the company.) distribute at least one type of safety educational material during the month, (This is an item such as a quiz, cross word puzzle, word search or other educational game.) and finally, hand out at least one Safety Awareness Item (SAI) per month, (A SAI is a tangible item that is given to employees to bring continued awareness to the monthly safety theme. For example, a Crunch candy bar can be used to remind co-workers not to be caught in a 'line of fire' crunch.). Check out a book called *The Untapped Secret for Selling Safety, And 401½ Tangible Items Guaranteed to Help Make That Sale*, to learn more about SAIs.

"You don't drown by falling in water." Motivational speaker and writer Zig Ziglar used to say, "You only drown if you stay there." Or, as I like to say, "When you are stuck, pull a lever." You may be interested to know that once the water dried and Otto and my father sorted out what happened they learned, as stated above, that the woman had turned the water on to make coffee. When she did so, the nozzle blew off of the faucet causing water to gush everywhere. In a panic, the women failed to realize the solution to the problem was right under her nose, she could have simply turned the water faucet off! In safety, a wall of water can sometimes push us back and put us on the sidelines. Remember, the solution is always there. The secret is to not stay in the water but to get up and find it. By the way, after turning off the water and helping mop up some of the mess. Otto and my dad repaired the nozzle with a 65-cent washer and were on their way. Don' be stuck on 'staying down.' Get up and pull a lever!

38 Leadership is . . .
The Hidden Traits of Leadership

When you hear the word 'leader' what figures from history come to mind? Many may think of George Washington and his courageous leadership in battle that ultimately gave birth to a new country. Some might picture Dr. Martin Luther King Jr.; relating to his dream of equality for all. Some might think of Abraham Lincoln, Nelson Mandela, Mother Teresa and the list can go on and on. But the one thing that each of these people have in common is heroic action for their cause. The problem with using these leaders as examples is that we often do not have the same type of revolutionary impact.

Instead, we fight bureaucracy, internal politics, habits and culture only to ink out small wins or minor adjustments. Our work can be tough, frustrating, thankless and seemingly unrewarding. Given this, maybe the best quote for leadership reads, "Courage doesn't always roar. Sometimes courage is the quiet voice at the end of the day saying, 'I will try again tomorrow.'" Sure, real leadership is George Washington, Dr. Martin Luther King Jr. and so on, but it is also knowing that you do your absolute best, push limits as far as they are allowed to be pushed and courageously show up the next day and do it again—that too can be great leadership. Below are three ways that you can be that courageous leader, the one who quietly, yet effectively, moves the middle, gets results experiences setbacks only to follow that quiet voice that says, "I will try again tomorrow."

Care—Randy was a tough, hard to get to know, son of a gun. He had more than twenty-five years of utility experience and didn't get along the best with the local management. I knew Randy and had actually worked with him when I was an apprentice. I was now the safety supervisor for the area. In this safety role I was responsible for nearly 400 utility line workers, substation workers and gas employees in out-state Missouri. I wanted to form a relationship with each person, showing them I cared, but given the miles between work sites, I couldn't see everyone in the course of a month. So, I got with Human Resources and received a list containing everyone's birthday. Then, on their special day I took a moment to write them a personal note—all 400, for two straight years. For Randy I said something like, "Happy Birthday, I really liked working with you back in the day. I always liked your funny stories. Work safe, Matt." I sent it and forgot about it. About five years later I happened to be in Randy's show up location and there on his locker with the pictures of his wife and kids was a faded piece of paper. I recognized it immediately; it was my note to him.

Family counselor and author Josh McDowell wrote, "Rules without relationship causes rebellion." If I needed to talk safety or enforce a rule with Randy, I'm confident he would have listened because he knew that I cared. It may not seem like moving mountains, 'charging the hill' or a decisive 'win' but showing someone in the organization that you care is true leadership. And, earning this trust, one person at a time will pay off in spades.

Keep Your Cool!—There were tense moments in the Cold War but maybe no period as intense as when Cuba declared itself a 'Soviet satellite State.' For the United States to have the much-feared communism only 90-miles from democracy was a nightmare come true.

The events of the cold war began as World War II ended. For a decade and a half, tensions between the two world superpowers grew, shifting to other countries, to Europe and one in the Caribbean, Cuba. In an effort to halt the movement in Cuba and protect the United States from a potential missile attack from Cuba, a small force of CIA trained Cuban exiles landed at the Bay of Pigs, Cuba in April of 1961. This effort, to over-throw Cuba's leader Fidel Castro, failed and only served to propel tensions between the US and the USSR to a new height. It was shortly after this, in 1962, that the world was saved.

At sea, using radar, a United States ship had detected a Russian submarine. The ship began launching depth charges in an attempt to bring the sub to the surface. In order to allude further detection, the Soviet submarine turned off all possible equipment including most fans and motors. While temperatures were soaring, so was the temper of the Soviet Submarine Commander. Finally, he had enough; he ordered a torpedo launch on the US ship. Soviet procedure required two additional officers to sign the order. One immediately signed. The second Captain however, refused. Instead he reminded the other two that naval procedures only allowed firing on another ship if a submarine's hull had been breached, which it had not. "He was a man who never lost his cool," a friend later said. The situation was tense. Most on the ship pressured him to sign, to launch an attack on the US Ship.

Safety can be tense. It doesn't matter if you are in front of a group of union workers explaining a rule change or a board of directors talking through injury statistics, the interactions can be intense. And, in these intense times leadership keeps its cool. Most historians strongly believe if a missile had been launched it may have started nuclear war. This Captain didn't know he was saving the world by a small act; doing what he considered the 'right' thing. And, in the moment, we don't realize we are potentially 'saving the day' by keeping our cool in tough situations. Save the world today, yes you can.

Be the Unfiltered Truth to the Top—After graduating from college, I was recruited to coach a 13 and 14-year-old AAU basketball team. Since I loved the game, I accepted. In two years of coaching, I'd like to say that my team was undefeated or that we won a title. In truth, we were just okay. My players learned the fundamentals and a couple of them went on to be outstanding high school players. For the record, I never threw a chair, recruited illegally or introduced human growth supplements. The highlight of those two years didn't come from my players. Instead, it was in the form of an invitation. Coach Bill, a local high school coach, invited me and a half dozen other coaches to attend a University of Missouri basketball practice.

Growing up a half-hour south of the Hearnes Arena, I was a huge Missouri Tiger fan and an even bigger Norm Stewart fan. Stewart was Mizzou's legendary Hall of Fame coach. In 32 seasons he led the Tigers to nine Big Eight Conference regular-season championships, six Big Eight Conference post-season tournaments, 16 NCAA Tournament appearances, including two Elite Eight appearances. Coach Steward holds a lifetime record of 634–333, which ranks 11th among Division I head coaches. In short, the man could coach, and being able to attend a closed practice was heaven for me!

Coach Stewart and his staff ran the players through numerous drills; dribbling, defense, rebounding, conditioning, and shooting then more defense. The coaches worked the players hard; it was intense. I noticed Coach Bill, sitting to my right, had brought a notebook and was taking notes. I figured he was jotting down some key drills to use with his team. Once I whispered a comment, a suggestion. Coach Bill jotted it down. I quietly wondered what he was doing . . .

The practice, and my dream day, finally wound down. The players were attempting their final free throws before hitting the showers, and Coach Bill tore off his notes page. He folded it and walked courtside. He shook Coach Stewart's hand, said something and gave him the folded sheet. As we walked out, I asked Coach Bill what the exchange was all about. He simply said, "I thanked the coach for letting us attend, and I gave him a critique sheet. Stewart really seemed to appreciate the notes," he added.

President Harry Truman picked up a nickname as 'Give'em hell Harry' but as he used to say, "I didn't give them hell, I just told the truth and they thought it was hell." One of the key roles a leader can play is being an unfiltered truth to the top. Doing this means that we first are close enough to where the work gets done to know what is happening there. Next, it means that we are able to work the organizational politics so that we can be heard, unfiltered, at the top. Unfiltered feedback, like that given to Stewart from Coach Bill is true leadership. In organizations, it is often 'unwanted' information that points to progress being slower than expected or culture not taking root as directed. Yet, if those running the organization don't know 'the unfiltered truth' then they cannot make critical decisions to correct potentially important issues.

Leadership is a tough, hard and mostly unrewarding task. We live for those rare occasions when leadership is charging through barriers and tumbling mountains. Between these career highlights, leadership is caring, keeping our cool and relaying information unfiltered to the top—consistently, over time. Leadership is, "the quiet voice at the end of the day saying, "I will try again tomorrow.""

39 The 5 Secrets of Successful Safety Teams
How to Make Common Sense Common Practice

"I have worked with all types of safety committees. At times a team will exhibit high performance. Then, in the same meeting, the team will reflect Milton Berle's comment and "keep minutes but lose hours?" At times, teams have been engaged and focused. Then again, they would just as easily be defined as "A committee is composed of people who individually can do nothing, who come together to conclude that nothing can be done" —Unknown

Finally, a team would show a glimpse of brilliance only to revert back and mirror this anonymous quote, "A committee is ten people doing the work of one."

The interesting thing about the comments above is that they are about the same team! Chances are that you have a safety team. Actually, chances are you have many safety teams within your organization. And, chances are those teams are greatly underperforming-and inconsistent! The Behavior-Based Safety revolution of the last couple of decades introduced us to the Safety Committee, a team of individuals gathering at regular intervals to do something. The problem is that the team, and management alike, are not always sure what the team should be doing nor how it should get done. At times, teams can do well and enjoy brief moments of success. But that's the problem; surrounding those rare insights are weeks and months of mediocrity. In general our teams seems to underachieve, month-in-and-month-out to the frustration of managers.

But, let's be fair to our teams, it's not all their fault. If we are completely honest, our best safety teams happen by chance and are the exception not the rule. In truth, under-achievement is a result of our failure to give our teams the key tools that would enable their successful. Employ these five Secrets for successful safety teams and revolutionize performance in the process.

Tip I—A Shared Belief System, SBS

Description: Any guess as to how many people are injured using table saws each year in the United States? It is over 60,000! In fact, in the time it takes to read this article, there will be another digit lost in a table saw related incident. To solve this safety issue most teams would examine the facts and conclude it is, 'just part of using the saw.' As a result, a team might set an aggressive goal to reduce injuries on table saws by 25%! Does that sound familiar?

A few years ago, however, a team had a different thought. In discussions around table saw injuries, one team member believed that these incidents could be prevented-and he was able to sell this belief to the entire team. "Understand, what we believe," says Max DePree, former CEO of Herman Miller Co., Inc. "precedes policy, procedure and practice." Armed with a shared belief system, this team dug deeper than any previous team. They discovered that if a small electric charge was placed on the blade, it could be connected to a meter. The meter continuously measured voltage and any voltage drop; a voltage drop would occur if a finger or hand touched it. When the meter detected a drop, it triggered a breaking mechanism that stopped the blade 10-times faster than a vehicle air bag is activated. The result was elimination of injuries. This result was only possible because the team believed in the same highly challenging yet achievable end.

Typical Practice: The elephant in the room that no one is talking about is the team's 'shared belief system.' The truth, a shared belief system is the foundation of a team's success. We can have a talented team, provide outstanding management support, follow through on the other four team secrets, hang high-glossy colored safety posters, etc., but if we fail to engage, understand and challenge the shared belief system, our team will not produce positive change or substantial and sustained results. The trap is failing to discuss beliefs and agreeing on a shared system, something like, 'all table-saw injuries are preventable.'

Common Sense Secret: Talk about the elephant! Hold honest discussions on individual beliefs regarding safety and injury prevention. Let these discussions lead to the establishment of a team belief statement. This is an overarching statement about what the team believes is possible in safety. Beginning with the words, "We believe . . . " can be the powerful starter your team needs for success.

Tip II—Crystal Clear Direction (CDC)

Description: "If you don't know where you are going, you'll end up somewhere else!" is a famous Yogi Berra quote with much rooted truth in safety teams. We would never hail a cab and tell the cab driver, "I'm not sure where I want to go, just drive." Instead, when we get in a cab, we have a direction, purpose, budget and an expected arrival time; after all, the meter is running! Unfortunately, too often we stick our safety teams in a cab and tell the driver to 'just drive.' We fail to give crystal clear direction (CDC). The best way to provide our teams with direction is through a specific end statement (Hackman). This is a statement of team destination but not the route (means) in which to get there. It is management's role to challenge the team with a CDC, it is the team's job to use the provided tools, training, knowledge and talent to get there.

Typical Practice: Management's job is to provide a specific end statement. The team's job is to outline their course of action based upon that end statement, or in other words, plan the travel route. The trap is that management, not wanting to interfere with the team, fails to examine and challenge the team's action plan. Just as a team should demand a specific end statement, an action plan to meet that end should be required of the team.

Common Sense Secret: Engage the entire team in order to meet the 'travel demands.' Generally safety teams have a few members with official titles such as chairperson and secretary. The remaining members however just show up and eat donuts. Once the specific end statement is communicated, have the team identify the jobs and roles needed for success. Assign each team member to one or more of these specific roles or tasks. This will engage each team member to a higher level. No more just showing up, with a specific role they will need to show up and be ready to participate.

Tip III—Clear Structure

Description: In December of 2006, Mid-Missouri was blanketed by nearly 24-inches of snow. Typically, six-inches of the white stuff can close schools; two-feet will immobilized the region. With the weight of the snow, many warehouses and factories were damaged as roofs collapsed. One of the more intriguing roof cave-ins however happened to a large horse barn. While most roofs failed in the first 24-hours, this one caved-in almost a week after the snow fell. Insurance officials investigated and determined that the roof was built to withstand the 24-inches of snow, when that snow was distributed evenly. There were, however, several periods of melting and re-freezing and the snow shifted, sliding and gathering along the mid-point on each side, causing failure.

"Every organization," Stephen Covey says, "is uniquely designed to exactly produce the results it achieves." Clear Structure can be defined as "what is acceptable on this team." It's the written and unwritten set of norms and rules that allow the team to, achieve, function well or cave-in under its own weight.

Typical Practice: Newly formed teams are eager and willing to share in the work of the team, thus teamwork and assignments (weight) are more-or-less distributed equally among members. Over time, if structure and rules are not clear, some team members will disengage and fail to complete assignments, roles and responsibilities. In the end, one or two team members will be carrying the weight of the entire team. Eventually, the team will collapse.

Common Sense Secret: Establish rules and norms and find fun ways to enforce them. For example, one team I served on had a rule that anyone late for a team meeting had to pay a $5 fine—all of the money going to charity. It was a fun way to enforce the team value of promptness . . . by the way, I was only fined once and was never late again after that!

Tip IV—Teams are Well Supported

Description: "An army runs on its stomach," is a saying used frequently and has a deeper meaning than just food. It refers to all of the supports including food but extending to fuel, clothing, medical, training, communications, transportation, logistics, etc., that are needed to make an military mission successful. In a similar way, our teams 'run on their stomachs.' Safety teams will flourish or succeed on the support they are given. Arguably the five key safety team supports are; time, budget, training/skill development, clerical/office assistance and feedback/recognition.

Typical Practice: Blaming lackluster team performance on 'the team.' At the end of the day, these are our teams. A team that is not performing up to par may be more of a reflection in what we have put in their 'stomach' than on the team. The trap is 'writing off' a team because of poor performance instead of asking tough questions about the supports that we failed to provide.

Common Sense Secret: Ask the team what they need! I realize it's a novel idea but too often we assume the team has what they need or they will ask if they don't. Instead, ask the team leaders key questions such as, 'How can I support you better?' or 'What keeps getting in your way?' Keep asking these questions until meaningful responses are received.

Tip V—Successful Teams are Well Coached

Description: In a safety world dominated by the behavior-based safety model, it should not be a surprise that coaching or feedback is a one of the five keys to a successful team. Today, there is a lot of pressure on safety staff and management alike to be great coaches. Well, forget about the pressure to be great and simply strive to be a 'GOOD' coach. GOOD means, <u>G</u>et in the game-one can only be effective if in touch with the pulse of the team. <u>O</u>ffer feedback-too often we observe a team and fail to offer any feedback, positive and constructive. Feedback is a must to keep the team on course. Be Optimistic-the team will feed off of your energy; remain up beat with positive energy. Finally, be Determined to make a difference-understand your role as a coach is to change the team when sometimes the team is striving to 'stay the same.'

Typical Practice: Failure to coach the team. Too often, we assume that since the team consists of a cross section of our leaders that they don't need our feedback . . . that assumption leads to ineffective teams. In truth, safety teams generally consist of field or floor employees. They are subject matter experts but not astute on the inner workings of successful teams.

Common Sense Secret: To ensure consistent and effective coaching, design a feedback sheet. The feedback sheet may contain questions that evaluate the specific end statement, action plan to reach that end, structure, support and coaching. Each meeting, a leader can fill-in the sheet. Afterwards, team leaders discuss the evaluation tool along with specific adjustments that need to be made to make the team more effective moving forward.

"It takes team work," John Maxwell said, "to make the dream work." In the end teams are a part of our business and safety culture. Making them an effective part of this process is the next step to our goals of a zero-injury workplace. Are your teams on target?

40 Is Your Safety Program FAIL FIRST?

Looking for Signs of a 'Fail First' Safety Culture

My wife, Stephanie, is a teacher. Although she has stayed home with our children for the last several years, she has still been involved in education. Over the last couple of years, she has mentored a young girl. Education for this girl, now nine, isn't easy. She has moderate to severe ADHD and SDI (Sensory Integration Disorder). Due to a stable home life, quality teachers and this added mentoring however, this girl has done well in school.

About two years ago, Stephanie began to notice a gap in the ability of this girl and her peers. This gap wasn't being reflected in grades, the girl was getting results, but it was troubling nonetheless. Stephanie had a strong sense that this child could receive additional support if she was tested for and qualified for special education services. The only problem was that this girl attended a smaller private school that did not have special education services. They would have to go to the local public school for a consultation.

Together, the teacher, parent and counselor met with the public school. They reviewed the medical diagnosis, psychologist's notes about the girl, current school performance and educational performance gaps. At the end of the consultation, the public school officials said that it appears this young girl could benefit from services but that her grades were too high. They explained that their program was a fail first program. Students had to first fail, which the girl had yet to do, before services could be provided.

Later that same night, Stephanie shared this story with me. I was incensed; how could a system allow failure before action? How could an organization let a young person fail, before providing a change, or additional help? The next day, as I worked with clients, I realized that our safety programs are sometimes fail first programs. Often we don't execute change; explore new programs, tools, technologies, procedures, etc., until there is first a failure, injury. Success, in education and in safety, seems to be about proactive identification and modification before failure. Below are signs that your safety system could operate in fail first mode.

Ineffective Incident Analysis—I see this all across the country. Someone gets hurt so the injured employee, supervisor and safety staff member will meet and review the incident. A report is generated with 'corrective actions.' The only problem, the corrective actions are only a recitation of the applicable safety rules and/or procedures. While the report is 'pretty.' with pictures and everything, it fails to do what is was designed to do, determine root causes and prevent reoccurrence! Can you say fail first? Unless the event is 'serious' with significant injury, loss of life or a major financial loss without an injury, these failures are allowed to continue, failure is allowed to be first.

Lack of Communication with Line Management—First line supervision is tough. A decade ago, when I was a first line in charge of electric crews, it seemed that everything stopped on my desk. When the company needed a new time reporting procedure, first lines did the training. When the construction cycle geared up, first lines were there. When an inspection procedure was added, first lines were accountable for the paper work. When . . . when . . . when . . . I could go on and on. The work pace and stress never slowed. That being said, this 'noise' takes us away from what is really important, listening and talking to our people. As a first line, I found that the comment or question that I ignored due to this other 'noise' always turned out to be a 'big deal.' It was a heads up or insight to a larger problem, but because I was too busy to listen, I was unknowingly fostering a fail first attitude. Are your first lines listening to the 'noise' or your people?

Near Miss Reporting—Okay, I don't want to scare you too much but here is the deal, if you have not received a near miss report in the last twenty working days, you are set up for fail first . . . scared yet? We all know the importance of near miss reporting. We all know that prevention before someone gets hurt is key. We all know that a near miss is just luck, a slight shift if timing or location, and the near miss incident could have been an injury. Knowing all of this, near miss incidents are still not reported. Really? What is driving this fail first attitude around near miss reporting?

What About the 'Small' Rules—Growing up in the utility business, working first as a lineman, then supervisor before serving on the safety staff, I realized there are a hand full of rules that are considered 'small' and 'optional.' It's stuff like setting out a wheel chock, after all, when is the last time a truck rolled away? Or, inspecting protective rubber equipment before every use. After all, it was inspected yesterday and was 'okay', why look at it again today? The problem, some of management team found these 'small' rules irrelevant as well. You and I both know that there are no small rules, and skipping any rule is deciding to fail . . . someday. What does your team think about 'all' of the rules and have you set some aside as 'small'?

Lack of Job Planning—From January through May of 2008, the tower erection industry had seven fatal events. The seventh occurred on May 22 in Florida as a young man named Joe Reed fell to his death. The National Association of Tower Erectors, or NATE took quick action as Don Doty, the acting Chairperson, pulled all 320 members together on a conference call shortly after Joe's death. Doty's message was clear. Hazards must be identified, communicated, then controlled and managed. The best way to do that is by identifying hazards through a tailgate or job planning session before work begins. "Setting the tone of the day with a safety meeting will keep safety procedures fresh in everyone's minds. We want to make sure everyone goes home safe at the end of the day, and 5 minutes could make all the difference," Doty said. Safety comes through planning or we fail first.

Coaching and Feedback—When I was a young electric distribution appren-tice working on a line crew, our supervisor had to perform a green card on us at least once per year. The green card was a job inspection, to ensure that we were performing work within the appropriate safety rules. I don't remember why they were called green cards except that they were printed on green paper. I recall the first green card that my supervisor performed. He pulled up on the job and sat in his truck. He made notes on the clipboard and after about half hour, he left. He didn't say a word about our safety performance, or lack there of, so a couple days later, I asked him. While he ensured me that everything was done by the rules there are at least two major things wrong with this. The first, once a year, and the second, no communication with the crew! This was nearly two decades ago and a lot has changed. We now understand that if we want to eliminate fail first, we need to have a proceduralized rotation for crew or job safety observa-tions with immediate and direct feedback to those performing the work. If we don't, we are in a fail first mode.

At the time of this writing, school is about to start. The before mentioned little girl, and here struggles to succeed, will start fourth grade. I strongly hope and pray that I can write a success story eight years from now, that this girl has graduated high school and is enrolled in college, yet without intervention we are afraid of what might have to be written. What will be written about your safety program this year . . . and in eight years? Will it be about the positive and proac-tive steps you took to eliminate a few 'fail first' symptoms or will it be about a series of injuries, some serious, and low results? Don't wait to fail, act today.

41 Safety, the Window to Operational Excellence

What Leaders Know About Accountability

Just about a year ago I was at a utility conference when a corporate safety director for a leading utility said, "If you can manage safety, you can manage any and every piece of your business. In fact, many investors are starting to look at incident and injury numbers along with the dividend payout and balance sheet. Wall Street understands that if you can manage safety, you can manage your business." Today, managing our business is more challenging than ever. As the economy continues to sputter, so does the demand for our product. Revenues may be down and we may be looking to make cuts and new ways to save dollars. A serious injury can lead to a seven figure cost. A soft tissue injury can lead to surgery and a mid-six figure cost. Safety represents a great opportunity for cost savings—not to mention it is the right thing to do. So what is the formula that can help us be even more effective in managing these safety numbers that Wall Street and investors are starting to look at? Well, some new research might surprise you.

A True 'Preventable' Tragedy—The cell phone rang; being in a meeting, I ignored it. It immediately rang again and I stepped out of the room. It was the regional dispatcher. I can still remember his words, "Electrical contact Matt . . . we've got two men down."

I told the dispatcher that I'd be there as soon as I could. I left the meeting and peeled out of the parking lot. I had an 80-mile drive to the work site; how could this have happened?

A recent study entitled, "The Peer Principle" by *Bloomberg BusinessWeek* published in May 2010 stated, "In the area of safety, our study found that 93% of employees say they see urgent risks to life and limb, and yet less than one-fourth of those who see concerns speak up about them. Rather, they wait for bosses or others to take action."

Once on site, I found that the crew had been setting poles and laying out phases to reconductor a piece of line. The six man crew, with over 100-years of experience between them, was going to work on the last pole then go home for the weekend. Given the experience of the crew and that fact that this job was normally done with three men, not six, it was a cake job for a Friday.

The only major hazard on the job was a 12,470 volt phase-to-phase overhead line. The crew knowingly crew positioned their truck under the line to avoid setting up on a busy road. Putting the truck there, under the only hazard on the job that could quickly end one's life, one would think they would have stopped and discussed this hazard; or, placed a spotter designated to watch the boom, making sure it stays out of the minimum approach distance, or ground the truck or cover the lines. They did none of these things, remember, "less than one-fourth of those who see concerns speak up about them." Shortly after starting work, the boom contacted the overhead line as the men were pulling material off the truck. Both receive an electrical contact.

How to Prevent This from Ever Happening Again!—One of the men, Brian, who received an electrical contact, was injured yet conscious and alert. The other man, Tony, was unconscious with shallow breathing. Emergency services were immediately called yet it was too late; the contact was fatal. Each day across the country there are, on average, 14 workplace deaths. These deaths represent a wife without a husband. A boy without someone to play catch. A beautiful young women on her wedding day without a dad to walk her down the aisle. In these cases, we are left to ask why, and how do we prevent this from ever happening again! Researchers may have stumbled on the answer by chance.

Researchers were studying organizations asking why there were differences in safety records. "We found," researchers later wrote, "That on the surface, the best and the rest looked quite similar. All were fastidious in keeping up with signage, inspections, compliance training, and enforcing safety policies. But we kept hearing unusual language in our interviews with the true standouts. It wasn't until we interviewed and surveyed 1,600 safety directors, managers, and employees that we realized we weren't really getting it.

"We were so focused on finding the key ingredient to building a perfect safety record that we missed the big picture our interviewees were trying to paint for us. That is, until one safety director used a bullhorn. Mike Wildfong, a general manager at TI Automotive, put it bluntly: 'You're missing the point. We lead in safety because we lead in accountability—not only as it relates to safety but as it relates to everything else we do.'"

Researchers thought that Mr. Wildfong might be right, but how would they prove it, and what does it mean to 'lead in accountability,' anyway? They rolled up their sleeves and looked deeper and what they found was pure gold! They later write, "Those supervisors and managers with the strongest safety records were five times more likely to be ranked in the top 20% of their peers in every other area of performance. They were 500% more likely to be stars in productivity and efficiency and employee satisfaction and quality, etc." It seemed that Wildfong was right but there was one last question, what did this 'accountability' look like?

"Since accountability appeared to be the key to safety as well as the full trove of corporate performance treasures, we then explored what made accountability tick in the leading teams and companies. Remarkably, cultures of accountability had little to do with bosses. Rather, it was all about peers."

In truth, the issue with Tony and Brian wasn't just that they violated safety rules. It was that there were six well trained and extremely qualified people on the job who clearly saw the hazard yet nobody gave feedback or said a word about the overhead wires. The research noted that when peers fail to give each other feedback, "It also explains what undermines safety in general. And similarly, it doesn't explain just what happens with safety; it also explains what happens with performance in general." Finally, the study concluded, "Peer accountability turned out to be the predictor of performance at every level and on every dimension of achievement. The differences between good companies and the best weren't that apparent when it came to bosses holding direct reports accountable. The differences become stark, however, when you examine how likely it is that a peer will deal with a concern."

The New Model for Safety Success!—The great statistician once said, "All models are wrong, yet some models are helpful." Just knowing that peer to peer accountability is the key to success doesn't help us improve that key skill within our companies. To that end, I wanted to offer a 'helpful' four step model, a path forward toward success. Remember, this new research clearly shows that top level safety performance is reached through Peer to Peer (P2P) accountability and this 'accountability' is the same as feedback! So, the goal is to employ a model that results in feedback driven accountability between peers.

Put it in a Procedure—We write procedures on everything, from rubber gloving to overtime rotation. Why wouldn't we write a procedure on what organizational accountability would look like and the steps it takes to get there? Within that procedure, we want to define, make a clear picture of, peer to peer feedback. We want to put measurement tools in place. We should consider including pay incentive tools in the procedure. Finally, we should integrate new hire evaluations to evaluate peer to peer feedback skill sets.

Practice It—In truth, we have for the most part developed a sit and listen culture, not a culture where peer to peer exchanges are not only encouraged but practiced on a daily or weekly basis. That best way to practice it is to infuse your current sit and listen activities like safety meetings, job planning sessions and company meetings with active exchanges, large and small group interactions and formal feedback times. The goal is to build this culture throughout the entire organization.

Pattern It—Most companies have followed the 'boss driven' form of accountability, and this has garnered some results. Companies have also placed a great deal of attention on first line supervisors, to 'hold employees accountable.' And, these practices can and probably should continue. However, given this research, the new role of a supervisor is to make sure his direct reports are 'patterning' peer to peer feedback. One supervisory skill that will become very important is the ability to evaluate if peer to peer feedback is happening or not, and the coaching skills to make it even more effective.

Pitch it!—After we have written a formal procedure, infused or companies with systematic ways to practice P2P feedback and patterned this concept for our organization, we should monitor for a year and when the organization has embraced this and become a firm habit, pitch the procedure. We are not building a bureaucracy of paper and procedure, we are working to keep guys like Brian and Tony safe—we want a true cultural change.

In closing, think about this old story. "A committee was appointed which consisted of four persons: Everybody, Somebody, Anybody and Nobody. A very important job came up, and the committee agreed that Everybody should be asked to do it. Everybody, however, was sure that Somebody would do it, and they all knew that Anybody could do it. But, alas, Nobody did it. Somebody got angry about that because it was Everybody's job, and Everybody thought Anybody could do it. But Nobody realized that Everybody wouldn't do it. So, it ended up that Everybody blamed Somebody when actually Nobody asked Anybody." Will Anybody start P2P feedback in your organization? Everybody can do it, maybe Somebody will do it, but it is our job to make sure it gets done, and Nobody gets hurt today!

Final Thoughts

I guess I have some good news and some bad news. The good news is that you finished the book—congratulations and thanks for reading. The bad news is that now that you have finished the book, your work has not ended, instead it has just begun!

There are three general phases of safety leadership development. The first phase is when one explores and captures safety leadership ideas and concepts—that is the purpose of this book. This phase, while requiring some work, is mostly about building energy, thinking and planning. In this phase, one gets to read and explore leadership books, concepts, materials and attend conferences to gather leading thoughts and determine what will work in his or her particular organization. This is the first stage, but unfortunately for most of us and most organizations, it is last stage.

"I never did anything worth doing by accident," Plato once said, "Nor did any of my inventions come by accident; they came by work," The word 'procedure,' not counting this section, appears 38 times in this book. In your organization, chances are that you have dozens if not hundreds of written procedures on all sorts of processes, both big and small. The bottom line is that important things get proceduralized. We have all agreed on the importance of safety leadership; that it is key to future success. Yet very few organizations have taken the second step in formal safety leadership development—having a written procedure on safety leadership. A procedure that encompasses leaders at all levels.

A formal safety leadership procedure takes all of those leadership concepts, ideas and thoughts and organizes them across all of your leaders. This group of leaders includes your CEO and senior staff, managers, supervisors and informal leaders. The procedure offers coordination between all groups and gives them a formal structure to align their efforts for overall better results.

The final step in formal leadership development is the implementation and ongoing evaluation of the procedure. How it will be evaluated and by whom, are good items to include in the plan.

Again, I hope this book has given you all sorts of ideas on safety leadership. Ideas you can take and put into action today! I also hope this book has given you the direction for long-term results. The notion that a formal safety leadership procedure not only makes sense, it is key to sustainable success.

Thanks again!
—*Matt*

Sources

"Navy doctor moves deep into firefight to save Marines," James C. Roberts, Stephens Media, LLC, 2011.

"On Call in Hell—A Doctor's Iraq War Story," Dr. Richard Jadick with Thomas Hayden, NAL Trade, 2007.

"On Call in Hell," *Newsweek*, Pat Wingert, March 20, 2006.

"Honor Culture" Linked to Accidental Deaths, Kim Caroll, August 15, 2011, American Broadcast Company.

Gladwell, Malcolm, *Outliers, The Story of Success*, Little, Brown and Company, 2008.

Leman, Kevin Dr., *What a Difference a Daddy Makes, The Indelible Imprint a Dad Leaves on His Daughter's Life.* Nashville; Thomas Nelson, Inc. 2000.

"New Study Finds Lifetime Costs of Injuries in Billions; Costs Associated with a Year Top $406 Billion" Center for Disease Control, April 2006.

McMillan, Alan C. "2007 CEOs Who 'Get It.'" *Safety & Health*, February 2007: 29–38.

Smith, Sandy. "ISP Columbus Takes Safety One Day at a Time." *Occupational Hazards,* October 2005.

Joe Stephenson, System Safety 2000, *A Practical Guide for Planning, Managing, and Conducting System Safety Programs.* 1991, Van Nostrand Reinhold, New York.

Gibson, Ken, *Unlock the Einstein.*

Buckingham, Marcus and Coffman, Curt, *Break All the Rules: What the World's Greatest Managers Do Differently.* Simon and Schuster, May 1999.

Mark Towers, "How to Deal with CAVE People: Citizens Against Virtually Everything," www.speakoutseminars.com.

Offerman, Lynne, Dr. *Harvard Business Review,* online version, January 2004.

SawStop, LLC, 9564 S.W. Tualatin Road, Tualatin, OR 97062, www.sawstop.com.

Hackman, Richard J. *Leading Teams: Setting the Stage for Great Performances,* Harvard Business School Publishing Corporation, 2002.

The Peer Principle, *Bloomberg BusinessWeek*—The Influential Leader, May 2010.

Book a Keynote Presentation or Safety Seminar

Matt offers safety keynote presentations and half-day, full-day and multi-day workshops on today's most urgent topics like employee motivation, leadership, supervisor effectiveness, accountability, organizational energy, and more. To learn more or book a presentation, log onto **www.safestrat.com** or call Matt (573) 999-7981.

Fun Client Feedback

"I felt that Matt was speaking directly to me through this book. It not only identified many of the issues facing my company's safety challenges but also many that apply to me personally. We oftentimes over-complicate safety. Matt provides a very simple recipe for success. He not only identifies key weaknesses that most of us face but he provides ideas and solutions that could be implemented tomorrow."

—Bill Dampf, CSP, Utility Safety Manager

"The layout of this book makes it very easy for a safety professional to pick up a story that is relative to what they are experiencing at the time and receive ideas that they can implement within their company. It is a quick read that develops the ideas with real world experience. I look forward to using this information as I our company continues its journey of re-culturing."

—Sheryl Wiser, Safety Manager, Fox Contractors

Book a Keynote Presentation or Safety Seminar

"I edit the writings of 100s of safety authors and Matt Forck is unique in his approach to safety leadership and saving lives. Matt makes safety reading easy with his humor, down to earth realism, and "from the frontlines" solutions. Matt has a sharp mind, big heart, and is "mission driven" to give safety professionals ideas and tools to build strong safety cultures, nurture strong effective safety leadership, and keep employees actively engaged in safety processes."

—Dave Johnson, Chief Editor (since 1980), Industrial Safety & Hygiene News (ISHN)

"This new book is another great thought provoking resource from Matt Forck for safety leaders at all levels . . . procedures, training and the technical aspects of safety are critical for sure, but nothing moves the needle on results more than getting leaders energized and engaged at all levels in a meaningful way. This book is full of winning ideas on how to do that effectively."

—Ken Bowman, VP Safety & Risk Control–Global Risk Management, ARAMARK Corporation

"We are all very busy, so if you want to take the time to read one book that will have the greatest impact on improving your safety program results, then you need to read What Safety Leaders Do.*"*

—Steve McKay, Willis Insurance Group

"Each year we look across the country and hand select the top 5% of speakers for our conference. Matt is one of those speakers!"

—Rick Donovan, Safety Professional. MO Mining Conference Planning Committee

"I just finished reading an article by you, in the August issue of Occupational Health & Safety, *titled 'ISMAs (Involved Safety Meeting Activities)' and it was excellent!"*

—Don Mersberger, Safety Coordinator, Johnsonville Sausage

"Very energetic and fun! Everyone should attend his seminar, the best of the conference!"

—ASSE Leadership Conference, March 2006

"A home run! Here's how to jump start your safety meetings. Inspiring, essential and a must for safety professionals."

—Mark "Tenacious" Towers

"You really hit a home run with the "rocks in the jar" demonstration and you'll be pleased to know it stayed on the podium the entire conference as a reminder to everyone of the great message you delivered. There was a tremendous amount of feedback during our debriefing session and several Operation's Managers had "take aways" about the "abilities" and making sure the big rocks/safety rocks were put in first. A great success in my book! Thanks again."

—Cindy DePrater, Turner Construction Company

"Absolutely remarkable at engaging his audience and keeping it lively!"

—ASSE Leadership Conference, March 2006

Other Books by the Author

You can learn more and order at **www.safestrat.com.**

Books to motivate you . . .

Check Up From the Neck Up
101 Ways to Get Your Head in the Game of Life

"Stink, Stank, Stunk!" describes the Griench, yet all too often these words reflect our thinking too. And how can it be otherwise as we are bombarded with negative news and under appreciated at work and at home? Yet, being stuck in the stink of life means that we are losing the best of ourselves; our dreams, passions, goals. To live our best life now, we need to become unstuck; getting our head in the game of life. To do that, we need a steady diet of strong, positive and powerful stories; short yet heartfelt, easy to read yet moving. In this book find blueberries for your brain. You'll get a check up from the neck up every time you open it and that is just what you need to have the life you want to live.

GUTSY
How to Go Until Time Stops You

"Life is difficult," is how renowned therapist and best-selling author M. Scott Peck opened his popular series, *The Road Less Traveled*. No disrespect to Dr. Peck, but one doesn't have to be a best-selling author to understand that life is tough. In truth, we are all looking for the same thing—to be GUTSY. Don't believe me, let's define it, "gut, adj, gutsy, -ier, -est: arising from within, from the innermost parts of the soul. Immediate and powerful impact, relevance, courage, brilliance, passion, fighter, significance." The problem is that in this 'life is difficult' world, it is easy to get derailed, pulled off the GUTSY track and into the daily grind of life. GUTSY will not only keep you focused and on track, it will remind you of just how special you are. Oregon State Director of Basketball operations, Coach John Saintignon said, "GUTSY is great, and I will use it over and over!"

Books to motivate safety!

The Total Safety Committee Checklist
A Step-by-Step Handbook for Safety Committee Success!
Don't you wish that safety committees came with instructions? After all, our committees are major investments in time and resources, yet they all too often fall short of expectations. Instructions, guidelines and tips for committees are scattered in procedures, sprinkled on the Internet, and scribbled on napkins . . . until now. The Total Safety Committee Checklist is your one-stop-shop for a month-by-month checklist of safety committee activities. Find both the basic 'must do' material for those committees that are just getting started, as well as extra credit items for highly effective committees. Safety committees finally have instructions!

ISMA (Involved Safety Meeting Activities)
101 Ways to Get Your People Involved
Let's be honest, attending a safety meeting is often like going to church . . . we are only there because we have to be! Add to that the fact that in a traditional sit and listen safety meeting, retention is as low as 10%; it makes us wonder why we even bother. But it doesn't have to be like that! This simple yet transformative book holds two key secrets. First, learn how to take any safety rule or procedure and transform it from sit and listen to get up and do; increasing retention four fold in the process. Next, the book contains 101 motivational involved safety meeting activities that will change your safety meeting culture forever. "A home run!" says Mark Towers, "Here's how to jump start your safety meetings. Inspiring, essential and a must for safety professionals."

Tailgate-101
Proven Stories to Begin Each Job Strong and Finish Safe!
In July, 2008, the tower industry was stunned by its seventh tragedy in as many months—this time a fatal fall in Florida. The National Association of Tower Erectors immediately called a 'stand down.' During a nation-wide conference call with over 320 of its members Don Doty, the Chairman, said, "Setting the tone of the day with a safety meeting will keep safety procedures fresh in

everyone's minds. We want to make sure everyone goes home safe at the end of the day, and five minutes could make all the difference." In other words-hold tailgates! Tailgates have been around forever, but how good are they? This innovative book provides proven tools to improve tailgates and safety meetings. Use these *101 Motivational Safety Stories* to enhance a tailgate, safety meeting or training session. These stories can be shared before a tailgate session to open the minds of the participants to the job hazards. Or, they can be shared at the conclusion of the meeting, to re-emphasize the importance of safety. Either way, they will help ensure the job starts strong and finishes safe!

The Untapped Secret to Selling Safety
And 401½ Tangible Items Guaranteed to Help Make That Sale!
How well do you sell safety? The truth is that we are at the mercy of our ability to sell, no matter how "tight" the presentation. Regardless of our education or the facts surrounding an issue, we are still in a position where we have to make the sale in order for a positive change to take place. And, the better we are at selling, the greater our results. The fact of the matter is that there are secrets to selling . . . even selling safety. One such previously untapped secret is revealed here and your safety results will never be the same! "Matt's passion for safety continues to shine through as he drives to inspire us to be the best we can be" wrote Bill Dampf, safety professional with three decades of experience. "Through this latest effort, he provides us with hundreds of ways to promote safety awareness to our employees. Although keeping our workers safe is always a challenge, this simple approach to helping us sell safety can be a tool that all of us can use."

Books on CD

The Everyday Secret for Results Everyday!
How many of you are sick-and-tired of being appreciated by your boss? How many of you have co-workers who complain about being over-appreciated? Okay, let's get real, how many have even heard what sounds remotely like a compliment in the last six months! Studies reveal that for every seven positive thoughts we have about another person we share just one. If we are going to lead

our organizations to top-tier industry performance, we will have to understand the purpose and power of feedback and in so doing, find 'The' Everyday Secret for Results Everyday! (30 Minutes)

Safety Duct Tape
Sticky Safety Pieces Guaranteed to Hold Your Safety Culture Together

There are a number of little factors that make for a strong and effective safety culture. We have leadership, belief systems, accountability and coaching, and feedback, just to name a few. In this fun, energetic, and enlightening audio CD, listen to live and studio recordings as Matt Forck, keynote speaker and trainer, offers tested and proven guidance to improve your safety culture— pieces guaranteed to make your organization even more effective! (72-minutes)

Meet Matt . . .
About the author

Matt Forck, CSP & JLW, is a writer, columnist and keynote speaker specializing in the fields of safety and human performance. Matt leads SafeStrat, LLC. SafeStrat, short for Safety Strategies, is dedicated to working with organizations to build sustainable safety strategies . . . for LIFE. (LIFE stands for Living Incident Free Everyday!)

When he is not traveling and working with clients, you can find him at his home in Columbia, Missouri. There, he enjoys the company of his wife Stephanie, coaching his son Nathan's basketball team and watching his daughter Natalie compete in gymnastics.

To learn more about Matt or book a keynote presentation, log onto **www.safestrat.com.**